Smallholders, Forest Management and Rural Development in the Amazon

The ongoing debate concerning the Amazon's crucial role in global climate and biodiversity is entirely dependent upon sustainable development in the region. Recognizing that forests are an integral part of the social fabric in the region, initiatives such as community forestry, small-scale tree plantations and agroforestry, as well as payments for environmental services, have aimed at conserving the natural forest landscape. At the same time these attempt to protect and enhance the well-being of poor local smallholders including indigenous groups, traditional communities and small farmers.

Against this background, this book analyzes numerous promising local tree and forest management initiatives taken by smallholders in the Bolivian, Brazilian, Ecuadorian and Peruvian Amazon to better understand the key success factors. The insights gained from more than 100 case studies analyzed by researchers from Latin America and Europe in cooperation with local stakeholders reveal the need for critical reflection on the initiatives targeting poor Amazonian families.

The book discusses an operational vision of rural development grounded on the effective use of smallholders' capacities to contribute to sustainable and equitable development of the region. It provides helpful information and ideas not only for scientists, but also for development organizations, decision makers and all who are interested in one of the major challenges facing the Amazon: to combine equitable development with the conservation of its unique ecosystems.

Benno Pokorny is an Assistant Professor at the Faculty of Environment and Natural Resources, Freiburg University, Germany. Previously he was a researcher at the Centre for International Forestry Research (CIFOR), based in Belem, Brazil.

The Earthscan Forest Library

This series brings together a wide collection of volumes addressing diverse aspects of forests and forestry and draws on a range of disciplinary perspectives. Titles cover the full range of forest science and include the biology, ecology, biodiversity, restoration, management (including silviculture and timber production), geography and environment (including climate change), socio-economics, anthropology, policy, law and governance. The series aims to demonstrate the important role of forests in nature, peoples' livelihoods and in contributing to broader sustainable development goals. It is aimed at undergraduate and postgraduate students, researchers, professionals, policy-makers and concerned members of civil society.

Series Editorial Advisers:
John L. Innes, Professor and Dean, Faculty of Forestry, University of British Columbia, Canada.
Markku Kanninen, Professor of Tropical Silviculture and Director, Viikki Tropical Resources Institute (VITRI), University of Helsinki, Finland.
John Parrotta, Research Program Leader for International Science Issues, US Forest Service – Research & Development, Arlington, Virginia, USA.
Jeffrey Sayer: Professor and Director, Development Practice Programme, School of Earth and Environmental Sciences, James Cook University, Australia, and Member, Independent Science and Partnership Council, CGIAR (Consultative Group on International Agricultural Research).

Recent Titles:
Evidence-based Conservation: Lessons from the Lower Mekong
Edited by Terry C. H. Sunderland, Jeffrey Sayer, Minh-Ha Hoang
Global Environmental Forest Policies: An International Comparison
Constance McDermott, Benjamin Cashore and Peter Kanowski
Monitoring Forest Biodiversity: Improving Conservation through Ecologically-Responsible Management
Toby Gardner, with a foreword by David Lindenmayer

Additional information on these and further titles can be found at
http://www.routledge.com/books/series/ECTEFL

Smallholders, Forest Management and Rural Development in the Amazon

Benno Pokorny

Routledge
Taylor & Francis Group
LONDON AND NEW YORK

earthscan
from Routledge

First published 2013 by Routledge

2 Park Square, Milton Park, Abingdon, Oxfordshire OX14 4RN
711 Third Avenue, New York, NY 10017

Routledge is an imprint of the Taylor & Francis Group, an informa business

First issued in paperback 2017

British Library Cataloguing in Publication Data
A catalogue record for this book is available from the British Library

Library of Congress Cataloging-in-Publication Data
Pokorny, Benno.
Smallholders, forest management, and rural development in the Amazon / Benno Pokorny. – First edition.
pages cm. – (The Earthscan forest library)
Includes bibliographical references and index.
1. Forest management–Amazon River Region. 2. Sustainable forestry–Amazon River Region. 3. Farms, Small–Amazon River Region. 4. Rural development–Amazon River Region. 5. Sustainable development–Amazon River Region. I. Title.
SD157.P65 2013
634.9'209811–dc23
2012036185

ISBN13: 978-0-415-66067-9 (hbk)
ISBN13: 978-1-138-57317-8 (pbk)

Typeset in Baskerville by Fakenham Prepress Solutions, Fakenham, Norfolk NR21 8NN

Contents

List of plates, figures and tables

Plates

The plate section appears between Chapters 4 and 5 (pages 106 and 107 of the text).

1 Amazonian smallholders represent a huge variety of ethnicities and cultures resulting from many indigenous groups and several waves of colonization: (A) shipibo in Peru (L. Hoch); (B) 'wild' settlers ('caboclos') in Brazil (B. Pokorny); (C) family immigrated from the Andean region in Peru (L. Hoch); (D) colonist from South Brazil (G. Medina); (E) descendants from rubber tappers in Bolivia (L. Hoch); and (F) descendants from Africans brought to Amazonia ('quilombola') in Brazil (G. Medina)

2 Study areas of the ForLive project

3 In many areas, rivers continue to be the principal mode of transportation for (A) production (C. Quette), (B) people (G. Medina) and (C) timber (J. Johnson), while in the advancing agricultural frontiers, roads are the principle mode of transportation. (D) Poor people using bicycles and motorbikes for transport (B. Pokorny), however heavier loads such as (E) charcoal (B. Pokorny) and (F) timber (G. Medina) require trucks.

4 Amazonian smallholders are engaged in a wide variety of agricultural activities. Typical traditional activities are the production of cassava (*Manihot esculenta*) (A) harvesting (J. Johnson), (B) processing (B. Pokorny) as well as (C) fishing (ProVarzea); more recent land uses include (D) cattle (L. Hoch), (E) perennials such as cocoa (*Theobroma cacao*) (L. Hoch) and (F) oil palm (Elaeis guineensis) (L. Hoch)

5 Examples of local forest management practices: (A) single-tree planting of mahogany (Swietenia macrophylla) in a cocoa (*Theobroma cacao*) plantation; (B) promoting regeneration of late pioneer species (e.g. *Pollalesta discolour*) in pastures in Ecuador and Brazil; (C) fruit production in home gardens; and (D) intensification of natural açaí (*Euterpe oleracea*) stands (photos by L. Hoch)

6 Many smallholders are engaged in the informal timber sector: (A) forest camp (B. Pokorny); (B) technically deficient cutting of

timber (R. Muehlsiegl); (C) transporting pre-processed boards with horses (L. Hoch); (D) local transport 'entrepreneurs' in Brazil (R. Muehlsiegl); (E) local saw mill (B. Pokorny); and (F) commercialization along the rivers (B. Pokorny)

7   For many indigenous and traditional families, the use of the wide range of non-timber forest products for subsidence and local commercialization is an important livelihood component: (A) uxi (*Endopleura uchi*) (G. Medina); (B) moriche palm (*Mauritia flexuosa*) (L. Hoch); (C) turu palm (*Oenocarpus bacaba*) (S. Ferreira); (D) Brazil nut (*Bertholletia excelsa*) (G. Medina); (E) rubber (*Hevea brasiliensis*) (G. Medina); (F) cashew (*Anacardium occidentale L.*) (R. Muehlsiegl); (G) cupuazu (*Theobroma grandiflorum*) (L. Hoch); (H) palm fiber (*Aphandra natalia*) (L. Hoch); (I) game (G. Medina); (J) batauá palm (*Oenocarpus bataua*) (L. Hoch); (K) honey (J. Johnson); and (L) charcoal (B. Pokorny)

8   The entire region is experiencing a strong urbanization dynamic: (A) traditional river house (B. Pokorny); (B) traditional river village (L. Hoch); (C) rapidly growing cities (Pucallpa in Peru) (B. Pokorny); (D) rural electrification remains often limited to the urban centres (L. Hoch); and (E) typical 'timber' city (Tailândia in North-eastern Brazil) (R. Muehlsiegl), Amazonian metropolis (Belém, capital of Para, Brazil) (R. Muehlsiegl)

9   In advancing frontiers, people are rapidly changing the landscapes: (A) area prepared for shifting cultivation in a forested landscape (G. Medina); (B) structured landscape created by small colonists (L. Hoch); (C) large cattle ranchers compete with smallholders on land (R. Muehlsiegl); (D) along the roads, forests are completely transformed in other land uses (R. Muehlsiegl); (E) yearly burning is expected to maintain the quality of pasture (R. Muehlsiegl); and (F) large areas are already free from trees and forests (L. Hoch)

10   Example of the landscape diversity created by smallholders (Medicilândia, Brazil) (adapted from Godar et al., 2008)

11   Region of Pucallpa (Peru) with areas dominated by indigenous people (right) and by colonists (left) (adapted from Godar et al., 2008)

12   Satellite images of part of the municipality of Medicilândia located along the Transamazon Highway in 1991 (left) and 2007 (right) (adapted from Godar, 2009)

13   The legal approach for large scale timber management by far exceeds smallholders' capacities: (A) Systematic tree inventories as central element of Reduced Impact Logging require well qualified teams (J. Souza); (B) tree cutting requires fast-working professionals (R. Muehlsiegl); (C) the use of skidders is only meaningful in large areas over long time periods (R. Muehlsiegl); (D) the purchase of specialized machinery for transport is incompatible with local realities (R. Muehlsiegl); (E) saw mills require professional management to be profitable (R. Muehlsiegl); and (F) mechanized timber harvest allows for attractive profits, but also requires large investments (R. Muehlsiegl)

14 Community forestry projects implemented by NGOs and governmental agencies with international funds can be found throughout the region: (A) sign indicating a community forestry project (G. Medina); (B) awareness raising and capacity building are typical NGO activities (J. Johnson); (C) promoted technologies often lack compatibility with the smallholders' reality (J. Godar); (D) portable sawmills are expected to increase smallholders' share of profits, but technical difficulties often arise (G. Medina); (E) transporting the timber out of the forest remains a challenge (G. Medina); and (F) despite significant volumes harvested, the profit margin is quite limited due to high production costs (G. Medina)

15 Transportation conditions are difficult in the Amazon, especially during the rainy season: (A) transport on rivers can be time consuming (B. Pokorny); (B) the majority of roads are constructed by private actors such as timber companies (B. Pokorny); (C) bituminized roads allow fast transport but require high maintenance costs (D. Martins); (D) many villages are cut off in the rainy season (L. Hoch); (E) heavy rain falls and high temperature make the roads highly sensitive to damages (L. Hoch); and (F) rivers represent a serious obstacle for long distance transport (C. Quette)

Figures

Tables

Foreword

This book is a rare example of the vast scientific literature on sustainable development in the Amazon as it combines research from different thematic and disciplinary perspectives with different geographic locations in order to draw practical conclusions on how to improve development prospects for smallholders in the Amazon. Therefore, the results presented in this book are not only extremely relevant for the future of the Amazon and for the international debate on global public policies and development, but also an example from which further research can definitely learn a lot.

The concept of sustainable development has been criticized a lot, by practitioners as much as by researchers who complain about the vagueness of the concept: How to translate it into meaningful public policies which produce serious advances in the environmental and social dimensions of development and do not subordinate them to economic stability and growth? How to conceive analytical and methodological research approaches which take into account the multiple dimensions and interlinkages of social, environmental and economic sustainability and still come to robust and general results about causal relationships?

Both questions remind us of assumptions and decisions which often remain implicit in politics and science, but which require an explicit discussion in order to avoid vagueness and make use of the innovative potential of the concept of sustainable development. Mainstream thinking in economics and political science is often based on the assumption that modern human society is independent from nature in the sense that advances in science and technology allow for the substitution and manipulation of natural resources and processes. Input and knowledge intensive agricultural production systems as applied today in Europe, for example, are taken as a model for 'modernizing' ways of using nature elsewhere, especially in the tropics.

This book questions this assumption in the sense that it first tries to understand the livelihood strategies of smallholders in the Amazon within their own rationality: smallholders are seen as rational agents whose practices are based on available resources and knowledge tested by experience. Any external intervention, therefore, has to pass the same test in order to produce lasting change. Otherwise they may lead to unexpected or no results at all. But the book also includes two further dimensions: the institutional setting to which smallholders' practices have to adjust, and the environmental conditions to which these practices respond and which they in part also shape. In this sense, the research presented in this book asks multidimensional questions about a practical problem (why it is so difficult to advance the quality of life of smallholders in the Amazon) and comes up with an answer which may surprise many: that smallholders have developed livelihood strategies which are well adapted to environmental conditions, and that the precariousness of their livelihood strategies is rather a result of dysfunctional institutional conditions and power relations.

This result requires policymakers to review their implicit assumptions when addressing issues of sustainable development in the Amazon, in the capitals of the Amazon states as well as in international organizations and other governments engaged in this endeavour through cooperation. Smallholders have the capacity to transform forest landscapes into a cultivated landscape that ensures environmental stability and contributes to their well-being – this is one of the strongest messages of this book, and it certainly is a provocation for many of the domestic and international public policy processes underway in the Amazon region.

Since the 1980s, the Amazon regions has been object of international debates focusing on the preservation of this biome, in order to protect its exceptionally rich biodiversity and its services for the earth's ecosystem, and to reduce carbon dioxide emissions from deforestation and forest degradation. This book makes three important contributions to this ongoing debate: First, it shows that family agriculture is based on an endogenous dynamic that pursues economic, social and environmental objectives in parallel and balances them out in the mid- and long term. Thus, family agriculture is advantageous for local development compared to large cattle farms and export crops. Second, although currently family agriculture merely survives in a precarious manner, it has an economic potential which can be unlocked when adverse context factors are changed (i.e. improving smallholders' organization and rural infrastructure, making technical assistance, loans and social services available and ensuring the rule of law). Third, external initiatives for the support of family agriculture have been well intentioned but generally failed from the point of view of their sustainability and replication. This is due to several reasons: the predominant view of external actors that current practices of family agriculture are deficient, their lack of knowledge on local conditions and past successful adaptations and the fact that most initiatives and projects focus on short-term results instead of lasting effects in the long term.

The book makes a fourth statement: the potential of family agriculture is threatened by the insistence of national and global public policies in integrating smallholders into international markets and to adjust them to the requirements of the globalized economy. There certainly is a fundamental incompatibility between a profit-maximizing market economy and a family agriculture, which aims at balancing work and leisure. But the globalized economy is currently shaken by the impacts of multiple crises: on financial markets, global warming, increasing food prices and peak oil. The international community is looking for new ways of dealing with these crises but few practical advances are made as the fundamental questions remain open: how to redefine and rebalance national and global interests, how to cooperate effectively in a world characterized by deep disparities in economic and political power as well as by rapid change in redistributing these disparities, as mirrored in the rise of China, India and Brazil.

What does all this mean for the search of a new equilibrium between economic growth, social justice and ecological sustainability? Formulated this way, the concept of sustainable development has a clear normative basis: these three objectives are not exchangeable, and the relationships between them are neither equal nor reversible. We can learn three lessons from this book: First, it is important to protect and promote alternative production systems, which are guided by securing a balance between economic, social and environmental objectives. Second, the ways in which external support is given and conceived have to change in order to strengthen the contribution of family agriculture in transforming the Amazon region into a cultural landscape, which combines socioeconomic and environmental objectives. Financial transfers within the climate regime need to be shaped in a way that they preserve and optimize the productive functions of smallholders. This would enable a sustainable development in the Amazon where the well-being of women and men in rural areas is not to the detriment of forests, and where integrated forms of land use replace the currently expanding pattern of land use in which environmental protection and productive use (cattle, soybeans, oil palms) are rather mutually exclusive.

Imme Scholz, Deputy Director of the German Development Institute/
Deutsches Institut für Entwicklungspolitik (DIE) in Bonn (www.die-gdi.de)

Acknowledgements

I want thank all the people who contributed to the ForLive project. Without the manifold inputs, support and assistance, this book would have not been possible.

To the local collaborators in the study areas in Bolivia, Brazil, Ecuador and Peru: Beyuma Salvatierra, Lurici Tirina, Salvatierra Linares, Vasques Chau, Chau Giese (comunidad Palmira), Yubanera Amapo, Amapo Yubanera, Yubanera Navi, Cepa Mayo and Tabo Macuapa (comunidad Buen Destino), Méndez Gutaica, Gualúza Ramírez, Monje Gonzáles, Duri Marigua and Adagua Tito (comunidad 12 de Octubre), García Cartagena (comunidad Buen Futuro), Francisco de Assis Monteiro, Milton Coutinho, Raimundo Rodrigues Xavier, Sr. Raimundão (Medicilândia), Maria Creusa Ribeiro, Cláudio Wilson Barbosa, Jomabá (José) Pinto Torres, Pedro Marciel (Porto de Moz), Badé and Delfim Oliveira Ferreira, Adamor Malcher and family (Prainha), Antônio Pereira (PAE Ecuador), Bruno Venturin, Rosecleide Leite, Antonio Reis Filho, José Omar Couto (km 107), Paulo Amorim, Ana Paula Santos Souza and the Fundación Vivir, Producir, Preservar (Altamira), Ranildo Moraes Viega and to the Consejo Popular of the Región Uruará (Santa Maria de Uruará), Giovanny Souza Guzzo (Anapú), Osvaldo de Oliveira (Rondônia), Agustin and Janneth Pisango (Pindo Mirador), Miguel Rigoberto Campoverde (Chinimbimi), Arsecio Kumpanam (Wachmas), Silvio Sandu and Verônica Pidru (Pajának), Julio Lojano Punin (Quinta cooperativa), Pablo Villegas, Juana Ihuaraqui, Malhy Murayari, Selmira Canayo, Cleydis Murayari, Norma Tamani, Roman Murayari, Elías Pinedo, Sr. José García (Caserío 7 de Junio), Roger Cumapa, José Murayari (Caserío Padre Bernardo), Anival Chávez, Enrique Dávila, Bernaldino Mahua, Vicente Inuma (Caserío San Juan), Luis Tuesta, Luis Alba Lostanau, Danika Carrión, Casilio Cumapa, Melgar Alvarado, Rosendo Cruzado, Salvador Rivas, Bonifacio Arcos, Wilildoro Hidalgo, Víctor Castro Lander (Neshuya Curimana), Pablo Silvano Barbarán and the families of the native communities of Callería, Preferida, Nuevo Saposoa, Patria Nueva and Panaillo, as well as the neighbouring villages.

To the researchers and assistants from the institutional partners of the ForLive project: Max Steinbrenner, Silvana Benassuly, Guillermina Cayres, Rodrigo Conduru, Paulo Contente, Fabricio Ferreira, Patrícia Freitas, Vanessa Gomez, Alice Luz, Átila Macedo, Deryck Pantoja Martins, Izildinha Miranda, Raimunda Monteiro, Socorro Oliveira, Mauro Rodrigues, Romy Sato, Daniela Sousa, Luciane Suarez

(Universidade Federal Rural da Amazônia – UFRA and the Fundação de Apoio à Investigação, Extensão e Ensino em Ciências Agrárias – FUNPEA); Armelinda Zonta, Oscar Llanque, Benjamín Añez, Huanger Ávila, Edgar Escalera, Mary Guevara, Karla Habu, Ofélia Landivar, Indira Monja, Carmelo Peralta, Pedro Pozo, Claribel Quitette, Delia Sangínez, Vincent Vos (Universidad Autónoma del Beni – UAB and the Instituto para e Hombre y Agricultura Ecológica – IPHAE); Stefan Gatter, Enma Arias, Rosa Masaquiza, Juan Pablo Merino, Marco Romero, Milton Reinoso, Fredi Tandazo, (Servicio Forestal Amazónico – SFA); Jorge Tizado, Javier Godar, Blanca Vizcaino, Raul Garcia, José María Gonzales, Vicente Manrique Simón (Universidad León); Freerk Wiersum, Bas Arts, Charlotte Bennecker, Chantal van Ham, Jessica de Koning, Doenja Kuiper, Rik Sools, Dirk Steenbergen, Chaves Villegas (University of Wageningen); Jürgen Bauhus, Benjamin Blum, Paulina Campos, Klaus-Dieter Düformantel, Ursula Eggert, Sebastian Hetsch, Lisa Hoch, James Johnson, Gabriel Medina, Inka Montero, Juan Carlos Montero, Sonia Ortiz, Marco Robles, Andrea Schäfer, Karol Trejo (University of Freiburg); Paulo Amaral, Fabio Bencid, Ivanilson Duarte, Rodolfo Gadelha, Zilma Nascimento, Lorenda Raiol, Rodney Reis, Suelen Renata, Marcio Sales, Elson Vidal (Insituto de Homen e Meio Ambiente da Amazônia – IMAZON); Jaime Nalvarte, Yolanda Ramírez, Danis del Aguila, Gladiys Campos, Juan Chávez, Susy Diaz, Juan Pablo Ferreyros, Augusto Figueroa, Iván Icochea, Hilario Murayari, Santiago Nunta, Miluska Palomares, Jorge Palomino, Jhan Pinedo, Roxana Ramos, Percy Recavarren, Carlos Sánchez, Pío Santiago, Raul Tafur, Pilar Yáñez (Asociación para la Investigación y el Desarrollo Integral – AIDER); César Sabogal, Bruce Campbell, Peter Cronkleton, Enrique Ibarra, Edith Johnson, Markku Kanninen, Jose Martinez, Julia Maturana, Robert Nasi, Katia de Oliveira, Pablo Pacheco, Rina, Gideon Suharyanto (Centre for International Forestry Research CIFOR); as well as to Cecília Alfaro, Jorge Israel Palomino Bullon, Tina Depzinski, Fernando Dick, Sabrina López, Roxana Ramos, Mariana Sánchez, Saira Saavedra, Cindy Schlicke, Jordi Surkin, Juan Carlos Torres y Jes Weigelt;

To our financiers and institutional collaborators of the International Congresses: European Commission – EC, Rights and Resources Institute – RRI, Capacity Building International – InWent, Government of the State of Pará, Serviço Florestal Brasileiro – SFB, Ministerio de Desarrollo Rural, Agropecuario y Medio Ambiente da Bolívia, Association of Amazonian Universities – UNAMAZ, Banco da Amazônia, United States Agency for International Development – USAID, Deutsche Gesellschaft für Internationale Zusammenarbeit – GIZ, Food and Agriculture Organization – FAO, World Wildlife Fund – WWF, Organización del Tratado de Cooperación Amazônica – OTCA, Instituto Brasileiro do Meio Ambiente e dos Recursos Naturais Renováveis – IBAMA, the Amazon Initiative, Grupo Nacional de Trabajo para la Participación – GNTP and the International Union of Forest Research Organizations – IUFRO.

To all the participants in the dissemination and discussion events organized and co-organized by ForLive, and those who helped me with reviewing and editing, especially to Kate Flick and Horst Thomas.

Acronyms

AIDER	Asociación para la Investigación y el Desarrollo Integral (NGO in Peru)
APPA	Asociación de Productores y Productores Agroforestales (smallholder organization in Bolivia)
ASL	Agrupaciones Sociales del Lugar (community land status in Bolivia)
BOLFOR II	Second phase of the Bolivia Sustainable Forest Management Project
BRIC	Countries at similar stages of newly advanced economic development, in particular referring to Brazil, Russia, India and China
CAIC	Cooperativa Agricola Integral Campesina (agricultural cooperative in Bolivia)
CDS	Comitê de Desenvolvimento Sustentável (Committee on Sustainable Development in Porto do Moz, Brazil)
CEJIS	Centro de Estudios Jurídicos e Investigación Social (indigenous support organization in Bolivia)
CFR	Casa Familia Rural in Brazil and Centros de Investigacion Agricola Local in Peru (Agricultural Family Schools)
CIAT	Centre for International Tropical Agriculture
CIDOB	Confederación de Pueblos Indígenas de Bolivia (indigenous organization in Bolivia)
CIFOR	Centre for International Forestry Research
CIP	International Potato Centre
CIRABO	Central Indígena de la Región Amazónica de Bolivia (indigenous organization in Bolivia)
CONAFLOR	Forests National Commission of Brazil
dbh	Diameter at breast height
DFID	United Kingdom Department for International Development
ETM+	Enhanced Thematic Mapper
FAO	Food and Agriculture Organization of the United Nations
FONCODES	Fondo de Cooperación para el Desarrollo Social (social development fund in Peru)

ForLive	Forest management by small farmers in the Amazon: an opportunity to enhance forest ecosystem stability and rural livelihood (EU financed international research project)
FSC	Forestry Stewardship Council
GEF	Global Environmental Facility
GNP	Gross National Product
GPS	Global Positioning System
GIZ	Gesellschaft für Internationale Zusammenarbeit GmbH
HDI	Human Development Index
IBAMA	Instituto Brasileiro do Meio Ambiente e dos Recursos Naturais Renováveis (environmental agency in Brazil)
IBGE	Instituto Brasileiro de Geografia e Estatística (Statistic agency of Brazil)
IDB	Interamerican Development Bank
IDEPRO	Instituto para el Desarrollo de la Pequeña Unidad Productiva (smallholder NGO in Bolivia)
IIRSA	Initiative for the Integration of Regional Infrastructure in South America
IMAZON	Instituto do Homem e Meio Ambiente da Amazônia (NGO in Brazil)
INCRA	National Institute of Settlement and Agriculture Reform
INRENA	Instituto Nacional de Recursos Naturales (environmental agency in Peru)
INRM	Integrated Natural Resource Management Approach
IPHAE	Instituto para e Hombre y Agricultura Ecológica (NGO in Bolivia)
ITERPA	Instituto de Terras do Pará (the land institute of the state of Pará, Brazil)
ITTO	International Tropical Timber Organization
MA	Millennium Assessment
MDGs	Millennium Development Goals
MINAG	Ministerio de Agricultura (Ministry of Agriculture of Peru)
MLA	Multidisciplinary Landscape Approach
MMA	Ministerio de Meioambiente (Ministry of Environment of Brazil)
NTFP	Non-timber forest products
PAE	Projetos de Assentamentos Agroextrativis (agro-extractive settlement projects in Brazil)
PAFSi	Planes de Aprovechamiento Forestal Simplificados (simplified forest management plans in Ecuador)
PCA	Principal component analysis
PENSAF	National Silvicultural Plan for Native Species and Agroforestry Systems in Brazil
PDS	Projeto de Desenvolvimento Sustentável (Sustainable Development Project)

PES	Payments for Environmental Services
PPG-7	Pilot Programme for the Conservation of the Amazon
PNFR	National Reforestation Plan in Ecuador
REDD+	Reducing Emissions from Deforestation and Degradation
RESEX	Reservas Extractivistas (extractive reserves in Brazil)
RIL	Reduced Impact Logging
SFM	Sustainable Forest Management
SLA	Sustainable Livelihood Approach
SNDCF	Sistema Nacional Descentralizado de Control Forestal (National System of Decentralised Forest Monitoring in Ecuador)
SNUC	Sistema Nacional de Unidades de Conservação da Natureza (national system of environmental conservation areas in Brazil)
STM	Sustainable Timber Management
SUDAM	Superintendence for the Development of the Amazon (in Brazil)
TCO	Terras Comunitarias de Origen (community land status in Bolivia)
TCL	Tratado de Libre Comercio (Free Trade Agreement)
TM	Landsat Thematic Mapper
UNCED	United Nations Conference of Environment and Development
UNDP	United Nations Development Programme
UNFCC	Unit Nations Framework Convention on Climate Change
USAID	Agency for International Development of the United States of America

Introductory note

Over the past few decades, considerable effort and resources have been invested in sustainable development in the rural Amazon region. In general, these efforts have aimed to conserve the natural forest landscape while also attempting to protect and enhance the well-being of local populations – particularly the immense diversity of indigenous groups and traditional communities living in the region (Sabogal et al., 2008b). These aims are relevant not only for local populations, but also globally, as demonstrated very clearly by the current debate on the Amazon's pivotal role in global climate change (Humphreys, 2008; UNFCC, 2011).

Development approaches have begun to change, reflecting the increasing awareness that the forest is an integral part of the social fabric in the Amazon region and that forest destruction has had a major impact on the rural poor. The combination of environmental protection efforts with poverty alleviation initiatives has become increasingly important. These combined endeavours may take the form of initiatives such as 'Community Forestry', small-scale tree plantation, agroforestry systems (Wunder, 2001) and, more recently, payments for environmental services in the framework of negotiations on initiatives for Reducing Emissions from Deforestation and Degradation (REDD+) (Engel et al., 2008). There are a number of highly interesting initiatives promoting more effective and sustainable use of natural resources in rural development. However, success has been rather modest in this area, as the destruction of forests is continuing with unabated speed, and the gap between urban rich and rural poor continues to increase (MA, 2005; UNDP, 2010).

With this social forestry development dynamic as the backdrop, a research group of nine partners from Latin America and Europe under the coordination of the University of Freiburg (Germany) launched, in February 2005, a European Commission-financed research project entitled 'Forest management by small farmers in the Amazon: An opportunity to enhance forest ecosystem stability and rural livelihood' (ForLive). By scanning and analyzing promising local initiatives for tree and forest management by small farmers, traditional communities and indigenous groups in the Bolivian, Brazilian, Ecuadorian and Peruvian Amazon, this project aimed to elucidate the reasons for the success or failure of such programmes from the smallholders' point of view as a basis for assessing the possibility of further applications in suitable contexts within the region (Pokorny, 2003).

Through the application of an innovative research approach combining traditional disciplinary research methods with participatory and transdisciplinary research, the project analyzed the local viability of the identified experiments and assessed their potential contribution to the smallholders' livelihoods in relation to their socioeconomic status and the environmental stability of the region. The research activities followed an iterative process of inductively generating, challenging and consolidating working hypotheses from empirical data collected in the field. These hypotheses were gradually formulated and discussed in stakeholder dialogues with an ascending level of comparative analysis, from case studies to the regional level. The different research processes were then synthesized into various coordinated activities. The analysis considered empirical data from five study regions, 17 in-depth case studies and more than 150 smallholder management experiments analyzed in complementary individual research projects conducted mainly by PhD and MSc students.

Over the four-year span of the project, nearly 100 researchers, doctoral candidates, Master's and other specialized students, field technicians, communities and families generated data and information that are not only useful for scientific purposes, but also highly relevant for development organizations, decision makers and the local families themselves (for more details, visit the project web site at http://www.waldbau.uni-freiburg.de/forlive). With the empirical evidence obtained in this project and in light of our gradually improving conceptual understanding, a series of events were organized in the four study countries, where more than 500 specialists, field technicians, decision makers, local families and their representatives, and students had the opportunity to reflect on the findings of ForLive. The participants discussed an operational vision for the development of the Amazon region that more effectively utilizes the immense social, economic and environmental potential of the people and resources in the region. Three other integrative events were held, providing an opportunity for information sharing and movement towards a change in development approaches in the Amazonian context: an international congress on rural development titled 'Provoking Change: Strategies to Promote Smallholders in the Amazon' and a discussion platform at the World Social Forum 2009, both held in Belém, Brazil, as well as an expert panel held during the World Forestry Congress IXX in Buenos Aires, Argentina, in October 2009.

The studies and discussions carried out within the framework of ForLive revealed that the current approach to development as employed by the majority of governmental and non-governmental initiatives targeting poor families in the Amazon is in critical need of revision. These studies further showed that very few of the initiatives – which were generally defined and designed by non-local actors such as governmental agencies, NGOs or experts – were successfully adopted or replicated by the families and communities. In addition, the project found evidence that these development initiatives even tend to accelerate, rather than slow, the processes of cultural deterioration and environmental degradation that have been observed in all study areas. It also became clear that the current legal-institutional framework still favours a development dynamic that largely ignores the social realities of local families and their capacities and demands and still disregards the integration of environmental considerations.

One of the reasons identified for this lack of success is that the perception of poor local families and communities and their socio-productive systems remains dominated by prejudices and generalizations that have become embedded in established approaches to rural development. However, such generalizations neglect these families' true capacity to participate in the development process. Classical development discourses still assume that traditional strategies for small-scale natural resource management are often indicted in resource degradation without significantly contributing to economic wealth (UNCED, 1987; Aldrich et al., 2006). In line with these assumptions regarding smallholders' limitations, the approaches of governments and NGOs and international approaches to environmental protection rely heavily on development strategies oriented towards export markets and the capacities of capitalized actors to apply agroindustrial technologies, to implement sophisticated forest management systems, to effectively organize the exploitation of minerals, oil and gas and to establish dams for the generation of hydropower (Pokorny et al., in press b). Interestingly, within the many projects focused on this type of *bigger/better* economic development, the capacity of local families and communities to contribute to the development of the region is ignored, even by the vast majority of organizations working for and with local people (Pokorny and Johnson, 2008b).

In contrast to the widespread neglect of the capacities of local people in both the theory and practice of rural development, the ForLive project revealed that the highly diverse socio-productive systems managed by Amazonian families do have an immense potential to contribute to the creation of environmentally stable landscapes that can serve as a foundation for robust economic development. Our studies further indicate that strengthening local capacities and perspectives as a model for an alternative approach to rural development that is more compatible with the human and environmental features found in the region will be essential for maintaining the cultural and intrinsically linked environmental diversity of the region. However, to utilize this local potential, small natural resource managers require high-quality and sufficient resources and adequate access to public services and markets because the feasibility of their economic activities relies on the payment of fair prices for the products and services offered to society (Vázquez-Barquero, 2002; De Schutter, 2011). Most importantly, these smallholders also require effective protection from large-scale capitalized actors interested in exploiting local resources, as they often misuse their competitive advantages in the global market.

Although these challenges are immense, this synthesis considers future directions for development from the lessons learned through ForLive. This work aims to promote a more promising trajectory for rural development that focuses on the elimination of the numerous current barriers to the strengthening of smallholders' socio-productive systems. Consequently, this approach moves away from the classical approach to development that focuses on the modernization of locally driven socio-productive systems in accordance with the capacities and interests of globalized markets. Against such a background, this work represents an attempt to reshape the discourse and practice of development in the Amazon, based on findings that have emerged from the many studies and dialogues that occurred during the four-year study period of ForLive. To structure the lessons learned and present the principal messages from this research, the book summarizes the extensive body of information by providing a review of the most important outcomes systematically supported by the analysis of case studies from the various study areas within the region. The text and presented examples are written in a way that hopefully facilitates a practical understanding of the results and assures the dissemination and comprehension of information among various levels of society. Thus, it is hoped that this book provides helpful information and ideas for people and organizations interested in one of the major challenges faced in the Amazon: to combine equitable development with the conservation of the unique ecosystems in the region.

Benno Pokorny, University of Freiburg, Germany

1 The ForLive project

The Neotropical rainforest represents one of the Earth's most complex and important ecosystems (Ghazoul and Sheil, 2010). Unfortunately, the rate of deforestation and forest degradation in this region is alarmingly high. Between 2000 and 2010, the net loss of forests in South America amounted to approximately 4 million ha per year (FAO, 2010). This loss was largely the result of national policies that mainly promoted agricultural development and colonization without properly considering the options for more sustainable management and conservation of natural forests. Deforestation in the Amazon is mainly caused by the owners of large cattle ranches and the agroindustrial and mining companies (Godar et al., 2012b). However, poor families, who currently own more than a third of the remaining forests (Sunderlin et al., 2008), exacerbate this process of continuous environmental degradation (Campos et al., 2001). Consequently, the promotion of sustainable development in rural areas of the Amazon has become increasingly popular. Sustainable development aims to achieve the dual goal of improving living conditions for rural populations while conserving ecosystems in the short and long terms (Prabhu et al., 1999).

Following the 1992 Earth Summit in Rio de Janeiro (UNCED, 1992), a wide range of public and private organizations began implementing initiatives that both fostered the sustainable use of Amazonian forests for economic development in the region and promoted the conservation of these ecosystems. In the beginning, this approach to Sustainable Forest Management (SFM), or, more precisely, Sustainable Timber Management (STM), was directed specifically at logging companies through the implementation of Reduced Impact Logging (RIL) methods in the commercial logging of timber and the targeting of viable international markets for Forestry Stewardship Council (FSC)-certified timber (Kaimowitz, 2003). Later, it became clear that reliance on commercial enterprises – which are the only actors with the appropriate financial and technical capacity to implement the proposed schemes for sustainable forest management – was not working as expected. Wider social issues were increasingly regarded as important, a trend reflected in the emerging discourse on sustainable development, self-regulation and global environmental governance (Arts and Buizer, 2009). Simultaneously, various social movements were initiated by indigenous people, rubber tappers and small farmers, who began to demand legal access rights to traditional lands and forest resources. This movement paved the way for new development approaches aimed at sustainable management of trees and forests by the smallholders themselves (Pokorny and Johnson, 2008a; Pokorny et al., 2009, 2010b), to provide local sources of income urgently needed by the inhabitants and motivate smallholders to value and conserve forests (Palm et al., 2005). These so-called 'Integrated Development Projects' (Wunder, 2001) included initiatives for community forest management, small plantations of profitable forest species, agroforestry systems, and, more recently, the establishment of payment mechanisms for environmental services (Pokorny et al., 2012).

The term 'smallholder' as used in this book refers to a broad category of actors including colonial pioneers, rural peasants, *quilombolas*, traditional river communities, indigenous groups and other social groups (see Plate 1). These groups represent a large cultural diversity but have in common the fact that they all utilize and cultivate natural resources, including trees and forests, as fundamental parts of their livelihoods and that they constitute a primarily family-based workforce.

Over the last decades, the situation of many inhabitants of the Amazon region has generally improved as a result of political reforms, development programmes and efforts from NGOs, often supported by the international community (Pokorny et al., in press a). However, in practice, despite several impressive examples of success, the efforts aimed at sustainable and equitable rural development remain far from reaching the goals of attaining long-term ecosystem conservation and establishing adequate living conditions for families in the rural Amazon (MA, 2005; UNDP, 2010). It remains unclear whether and under what conditions the stated objectives of current development strategies can be met. Although governments and NGOs are continuing with classical development approaches combining economic development and the integration of the local poor in global value chains, fallacies in the assumptions inherent in this framework have been noted (De Schutter, 2011; Pokorny et al., 2012).

Our experience has revealed that the experts and smallholders contacted during ForLive almost exclusively considered the most promising natural resource management initiatives to be those that depended on (often significant) external support, whereas the ways in which smallholders managed agricultural crops, trees and forests were generally perceived as ineffective and founded on obsolete knowledge and technologies. This perception, in combination with preconceptions regarding smallholders' lack of sensitivity, their supposed incapacity to use scientific knowledge and advanced technologies and the assumed lack of proper organization schemes, have generally been suggested as the root causes of the processes of deforestation and environmental degradation.

This latent lack of appreciation of smallholders' socio-productive models, not only by external actors, but also by the majority of the families themselves, marginalizes smallholders in the discussion of strategies for sound rural development in the Amazon. Existing development initiatives encourage smallholders to adopt development approaches that are unstable and yield mixed results in environmental conservation and rural poverty alleviation. With this in mind, the following chapters systematically challenge prevailing assumptions in the development field about smallholders and their land-use systems by reflecting critically on how they use their natural resources, especially trees and forests (Chapter 2), the potential contribution of their socio-productive systems to rural development (Chapter 3), the reasons for the marginalization of their cultures and socio-productive systems (Chapter 4), the support available to them (Chapter 5), the implications of this support (Chapter 6), barriers impeding the mobilization of their potential (Chapter 7) and the conclusions that can be drawn from these findings (Chapter 8). As a first step, this introductory chapter explains in detail the analytical approach, the research organization, the methods applied and the analysis of the case studies from study areas in Bolivia, Brazil, Ecuador and Peru.

Methodology and research principles

The ForLive project first identified forest use in the Bolivian, Ecuadorian, Brazilian and Peruvian Amazon by small farmers, traditional communities and indigenous groups. Promising initiatives with the potential to result in sustainable use of trees and forests were further investigated to determine the reasons for their success, forming a basis to assess opportunities for the application of these initiatives in suitable contexts elsewhere in the Amazon (Pokorny, 2003). By applying both traditional research methods and participatory and transdisciplinary research methods, the project aimed to analyze the social, environmental, technical, institutional and financial aspects of the identified case studies to verify their local viability and determine their potential contribution to smallholders' livelihoods in relation to the socio-economic status of smallholders and the environmental stability of the region.

Another objective was to reflect critically on the range of viable options for supporting smallholders facing the challenge of preserving, securing and utilizing their resources effectively in a highly dynamic regional context. The initiatives identified were adopted as reference points for understanding and discussing the existing options, thus forming a basis for finding new potential options to improve the socio-economic situation of a larger number of smallholders in the region. In this sense, the project searched for ways to resolve one of the primary concerns in the region: how to combine conservation of the forest ecosystem with rural development for the benefit of the local population.

The theoretical understanding of the realities found in the case study and study areas relied on the actor-network approach. Archer's Morphogenetic Approach (Archer, 1995; Hodgson, 2000; Elder-Vass, 2007), which assumes that there is interdependence between the *structures* of a society and *actors* within that society, was used. In this sense, the project adopted a dialectical perspective to structure and agency – recognizing the agency of the actors in elaborating and changing previous structures to form new ones while also recognizing the role of *antecedent* structures in influencing actors' actions. This approach acknowledges a strong effect of social structures on the individual scope of action without denying actors' capacity to overcome this effect under specific conditions (Weigelt, 2011). Consequently, the analysis did not perceive smallholders as simple 'end-of-the-line' actors implementing policy decisions made at higher organizational levels and who were thereby totally dependent on higher organizational structures. Rather, smallholders were observed as being active players, with a specific social identity emerging from environmental, social and economic contexts. Although the structures in place strongly influence smallholders' social and production systems, smallholders can also reshape and transform reality within the boundaries of the existing institutional structures and ultimately, to a certain degree, potentially redefine them (Cleaver, 2002).

In accord with this theoretical understanding, the project pursued the principles of the Integrated Natural Resource Management Approach (INRM) outlined by Lal et al. (2001) and Sayer and Campbell (2001) for generating information on the topics of analysis:

Integrating spatial and temporal scales: Our research considered processes affecting land-use dynamics and smallholders' management decisions as occurring in different time frames and on different spatial scales. Thus, the research design distinguished parameters that change slowly and have long-term effects from parameters with the potential for abrupt dynamic changes. We examined the overall context of the case studies analyzed to develop an understanding of the different scales of operation of smallholders' social production systems and the associated economic, social and environmental interactions. Thus, an understanding of the cultural, political, environmental and economic aspects of smallholders' socio-productive systems was attained at the household level, through case studies, and in the broader regional context of the study areas. By comparing information across these scales, situated in different contexts typical of the region, we sought to ensure a sound basis for extrapolation and interpretation of the wider implications and long-term relevance of the findings. Through this continually expanding information base, we intended to capture the social and environmental diversity in the region while linking the various scales of space and time.

Integrating multiple disciplines: ForLive recognized that the reality of smallholder socio-productive systems cannot be understood within clear-cut disciplinary divisions. For example, forest disturbances and regrowth dynamics are greatly affected by interactions between people and the natural environment. A holistic approach was chosen to obtain valid results on the viability of natural resource management systems that also meet social needs. This approach combined ecological and socio-economic assessments, particularly when considering the objectives of the smallholders and their degree of access to resources. Thus, in the analysis of each case study, the social, cultural, political, technical, economic and environmental parameters were integrated to identify and understand their possible interrelations and dependencies.

Multi-stakeholder collaboration: To increase the relevance and quality of the project findings and enhance the use of the generated knowledge, the project systematically explored opportunities to stimulate local learning processes and to actively engage different groups in the research process. In particular, we drew on relevant governmental and non-governmental organizations with prolonged involvement in the study area and often directly involved in the support of the selected case studies. Most importantly, the project invested in actively involving the smallholders themselves. In addition to communicating about the ongoing research to promote local understanding and transmit scientific knowledge, the project also provided families and communities with opportunities to create their own research agenda based on their interests and capacities. In each study region, the project established communication networks with the relevant local and regional actors to exchange information and promote discussions on the topics and findings of the projects and local research agendas, with researchers acting as facilitators of learning processes at the local and regional levels (Pokorny and Johnson, 2008c). The involvement of students to help bridge different worldviews was found to constitute an important part of this process.

The research team adopted a comparative case study approach (Yin, 2009), aiming to reflect critically and consolidate flexible working hypotheses. This process was based on a review of prior scientific knowledge together with results from the emerging empirical research. The systematic analysis of a wide range of case studies and contexts with a common set of parameters allowed us to draw general conclusions about the topics analyzed. The research included empirical data from five study regions and 17 in-depth case studies, as well as more than 150 case studies from individual, complementary research projects conducted mainly by PhD and MSc students.

Field research for the in-depth case studies and study regions used a standardized set of parameters and methods defined in three work packages addressing the (1) institutional framework; (2) local livelihoods; and (3) environmental aspects of local natural resource management. Furthermore, the case studies analyzed in the individual research projects provided additional data that were considered for comparative analysis.

Research activities were guided by an iterative process whereby working hypotheses derived from field data were generated inductively, challenged and then consolidated. These hypotheses were formulated gradually in dialogues with stakeholders, beginning at the local level and continuing to the regional level, after a comparative analysis of results. The different research activities were then brought together and synthesized into four coordinated activities: (1) a comparative analysis of the standardized data sets generated in the case studies in accordance with the specific objectives of the institutional, livelihood and environmental work packages; (2) the establishment of internal project mechanisms for exchanging and reflecting on individual findings and insights gained from the individual research projects; (3) a systematic assessment of emerging working hypotheses in all relevant contexts; and (4) a sequence of public seminars and conferences to consolidate the results generated and conclusions drawn for the formulation of relevant and appropriate general directions, opportunities and means to define a new approach to development in the Amazon region and beyond. Nearly 500 experts, researchers, policy makers and representatives of NGOs and smallholders participated in this activity.

Nevertheless, it must be noted that the sample for this research – although diverse and highly relevant for the region – was selected subjectively; hence, this analysis represents the specific realities and experiences of this stakeholder sample. In this study, the focus was placed on selected promising initiatives, meaning that the sample lies at the more successful end of the range of experiments in the region with respect to social, economic and environmental concerns. This selection implies that the majority of the results presented herein cannot be underpinned by a statistical analysis. It is probable that the vast majority of families and communities in the Amazon are confronted with worse conditions, with fewer benefits from support initiatives, and are under greater threat from the frontier politics of capitalist actors and globalization.

Analytical framework

To conduct an effective analysis of the selected experiments and contexts of smallholder management of trees and forests in the Amazon, a common analytical framework was designed that matched the methodological approaches adopted in the three analytical areas: institutions, livelihoods and the environment. Research in each dimension was guided by a set of specific research questions developed in a dialogue between the ForLive researchers and relevant groups of actors in the study regions and on the basis of an extensive literature review. This process included a sequence of workshops and bilateral meetings at the beginning of the project. The analytical framework was then refined and translated into a standardized methodological approach, defining the parameters and methods for field research. This section presents the essential elements of the analytical framework, which are described in more detail in the work package descriptions. The descriptions of the three work packages by Wiersum (2007), Vos et al. (2007) and Godar (2007) are available at the ForLive homepage (http://www.waldbau.uni-freiburg.de/forlive).

Research on institutions

Within the scope of ForLive, institutions were understood broadly as constellations of normative interactions and processes guiding human activities that may be formal, based on rules or laws, or informal (socially embedded, in the form of unwritten codes and rules). In this sense, ForLive considered institutions as 'the rules of the game' that people and organizations follow. An organization was defined as a structured group of people bound together by several common purposes to achieve particular objectives (Scott, 2001). In accordance with this broad concept of institutions, institutional research in ForLive aimed to improve our understanding of the structure and influence of institutions on smallholders' decision-making processes regarding the use of natural resources. This approach also involved the identification and description of relevant institutions, followed by an analysis of related trends and processes.

The research explicitly considered two main forces that, to a large extent, determine the scope of action of smallholders: first, historical power relationships (still currently embedded in institutions) accentuating social inequalities; and second, an ongoing process of institutional transformation arising from globalization, which is dramatically changing the relatively simple institutional arrangements for natural resource management in the Amazon. Although these forces have been perceived as enormous barriers to establishing robust institutional arrangements for the effective management of resources on a local scale (Ostrom, 1999; Becker and Ostrom, 1995; McKean, 2000), they also provide a lens through which we can gain an understanding of smallholders' decisions as part of a process of transformation and dynamic development. To this end, three main trends were found to challenge the still-existing historical power structures: (1) the gradual opening up of the Amazon region to more settlers, companies and institutions as a result of their inclusion in changing government and marketing networks; (2) the discourses and related governmental attempts to establish policy arrangements expressing broader societal perspectives and thereby enabling greater influence by regional, national and international civil organizations (decentralization and devolution); and (3), the gradual development of international treaties for the preservation of the Amazonian forests and access to international assistance using international standards to achieve this goal.

In view of this dynamic multi-scale context, an analytical framework was designed to learn how smallholders were responding to this changing institutional reality, how power structures affected transition processes, and whether and to what extent smallholders themselves were influencing institutional arrangements (see Figure 1.1). This analytical model assumes that there are critical points of intersection between individuals' scope for action and external institutions. For analytical purposes, the identification of existing divergences between local and external institutions was most crucial for the characterization of differences in normative values and social interests between these levels, which also indicated structural discontinuities (Long and Van der Ploeg, 1989). The research on institutions concentrated on analyzsing these interfaces to better understand the nature and origin of the ongoing transformation processes and their influence on the local institutional conditions shaping the scope for action in smallholders' decisions regarding their socio-productive systems. Thus, institutional research analyzed the factors shaping the location-specific development and changes in the social and productive systems of smallholders within the intersections of local realities and external institutional conditions.

Special attention was given to understanding smallholders' responses in this transitional context by applying the concept of 'institutional bricolage' (Cleaver, 2002). This concept describes the relationship between the actor and institutions as a process whereby the actor uses certain institutional factors that create both opportunities and limitations in the choices available to the actor. Thus, the institutional research aimed to assess the scope of action of smallholders as dictated by external conditions and to investigate the extent to which smallholders' decisions stemmed from individual agency, from rules embedded in joint social action or

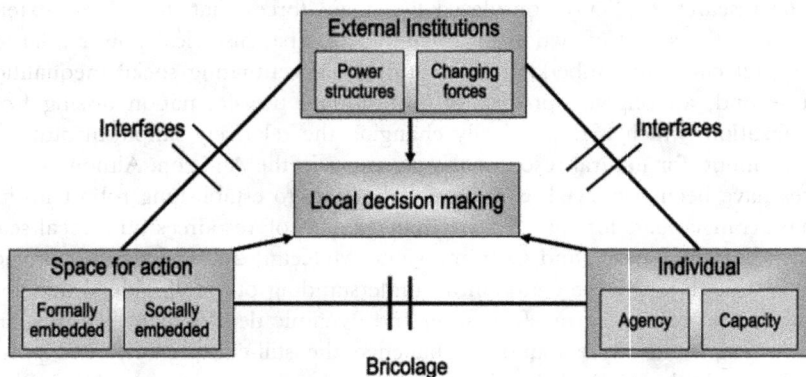

Figure 1.1   Analytical framework for analyzing institutional factors (adapted from Wiersum, 2007)

from compliance with externally defined norms and laws (De Koning, 2011). In view of recent efforts to establish regulatory frameworks that better consider local realities (see Chapter 5), special attention was given to studying the influence of these frameworks on smallholders' decisions on natural resources.

In the field, multiple qualitative and quantitative methods were used to collect data. Detailed qualitative data were collected in case studies through in-depth and semi-structured interviews with individual key respondents and groups. Field observations were also recorded during this process. The information gathered enabled a preliminarily description of production systems and social organization and basic community features. Participatory group research was undertaken with the families. Venn diagrams illustrating the local institutional settings were created and, based on the insights gained, a follow-up survey questionnaire was developed and given to 36 households throughout the different study regions.

For each case study, the institutional context conditions were analyzed. Interviews with key informants were structured on the basis of a review of the scientific and grey literature. Special emphasis was given to understanding the historical processes that shaped the existing power structures, current land-use dynamics and ongoing transition processes. Information was consolidated in a process of triangulation using participant observation and saturation sampling techniques, whereby the scope and size of the sample were adjusted in accordance with the categories, themes or explanations emerging as the study progressed. Thus, a flexible, iterative approach was applied to sampling, data collection, analysis and interpretation (Marshall, 1996).

An inductive approach was used in the data analysis: data were organized into categories and the relationships and processes were analyzed. Quantitative data were also analyzed with standard statistical tests. Findings from the case studies were further synthesized into key messages through a content and domain analysis followed by a final search for patterns between different smallholder forest management systems and institutional factors and processes.

Research on local livelihoods

For the analysis of local livelihoods, the project adopted the concept introduced by Ellis (2000) and principally built on the work of Chambers and Conway (1992): 'a livelihood comprises the (natural, physical, human, financial and social) assets, the activities and the access to these (mediated by institutions and social relations) that together determine the living gained by the individual or household'. Thus, the understanding of livelihoods adopted by the project was not restricted to income-related activities, but also included issues such as social security, stability, vulnerability, trust and identity. These interactions are embedded in a contextual framework that is deeply affected by rapidly advancing agricultural frontiers and the opportunities and threats accompanying this dynamic (see Chapter 4). Thus, in accordance with the conceptual approach to institutional research outlined in the last section, local livelihoods were perceived to be embedded in the social, cultural and historical conditions in which the smallholders live, produce and reproduce (Kaag et al., 2004). However, the smallholders, as agents, also have the power to act and challenge or change the rules for controlling, using and transforming assets (Bebbington, 1999; Schmink, 1984). This concept includes the classical perspectives of the household unit, the community, the production sector and the commodity chain (FAO, 1992), and encompasses a wide range of internal, external and interactive interpersonal networks that embrace a variety of activities, particularly gender-specific differences.

Our research focused on the livelihood strategies of smallholders, that is, the combination of activities people choose to perform to transform their assets and achieve well-being in a specific context (Chambers and Conway, 1992). These strategies include factors such as contingency plans in the face of unforeseen events or adverse pressures and handling risk, organizing work, controlling income, consumption and investing in well-being. For the empirical analysis of the ForLive case studies, we adapted the Sustainable Livelihood Approach (SLA), which was developed to facilitate a holistic and flexible analysis of local livelihoods from the perspective of the smallholders themselves. This systematic analysis of the relationships and dynamics of the various factors that affect opportunities for families living in poverty promised to reflect local realities and needs more adequately than previous analyses (Figure 1.2) (DFID, 1999; Bebbington, 1999; Farrington et al., 1999). In addition to the conventional analysis of smallholders' capacities, activities and capital (human, social, natural, physical and financial), we wanted to gain a sound understanding of smallholders' options for (and limitations to) improving their situation in response to emerging opportunities and threats.

Consequently, the sustainable livelihood framework that we derived placed emphasis on two essential aspects: first, the understanding of the role of each of the livelihood components in the local socio-productive system, particularly in relation to the role of forests and related goods and services; and second, an assessment of the opportunities for restructuring or reorganizing these

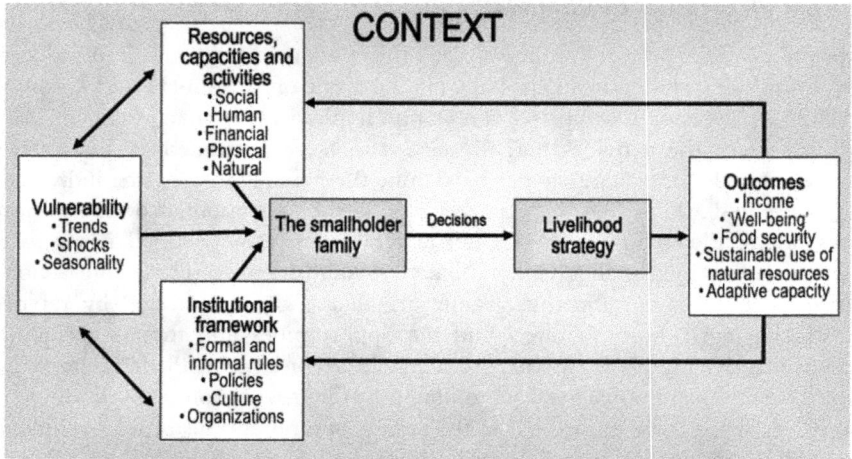

Figure 1.2   Conceptual framework applied to study local livelihoods

components to enhance the level of well-being and the quality and stability of local livelihoods. In addition to gaining a multidimensional understanding of local living conditions and the strategies developed for production, the analysis also focused on vulnerability and adaptive capacity as key issues. This included an assessment of the smallholders' ability to cope with and recover from stress or adversity and thereby maintain or expand their capabilities and assets. Uncertainty regarding the context of stability was accounted for in the analysis by considering two situations: regular, foreseeable conditions and – the much more frequent situation – conditions of insecurity and drastic change.

A special branch of the livelihood research was dedicated to the role that natural resources, particularly forests, play and could potentially play in the household economy of smallholder families. This branch of the research included the identification and quantification of key products and services derived from forests in local management schemes and the exploration of the perceptions of local families regarding forests and their motivations for managing them. The analysis also considered the economic potential of forest products at the household level and their potential contribution to regional welfare. In each study region, information was gathered on timber and non-timber forest products (NTFPs), their utilization and their market prices.

Data were collected using a graduated systematic approach in the selected case studies. In each case, a representative range of smallholders was sampled, taking into account their specific economic and cultural links to trees, forests and their products. Data gathering, including semi-structured and open interviews, partici- patory observation and an extensive review of the literature, was performed at three levels: (1) *the household level* of the selected families; (2) *the community level*, representing the array of families living in an area under similar conditions; and (3) *the context level*, which significantly overlapped with the information gathered for the institutional analysis. Whenever possible, quantitative data were collected in the field to create a sound empirical basis for analysis. However, the livelihood assessment relied primarily on the analysis of information from interviews and workshops with local experts and different actor groups, as well as field observations by the researchers.

In a preliminary step, a survey was conducted in each study area to generate basic data on households and social settings, the assets in particular forests, production systems and management practices. This involved gathering quantitative data on smallholders' properties and their various production components. Based on this information, a set of criteria was developed to describe and classify livelihood strategies; these criteria were then systematically applied in comparative analysis and data interpretation. After the components of the smallholders' production systems were identified in the case studies, management practices and means of processing, transport and commercialization were described and analyzed to identify related inputs and outputs. Together with the quantitative information on volumes, work-force requirements and prices, these descriptions served as the basis for a financial analysis of the different production components (Medina and Pokorny, 2008, 2011). The data were then used to examine the potential determinants of the diversified income portfolios of rural smallholders (Reardon et al., 1992) and to examine the relationships between income, household characteristics and barriers enabling smallholders to enter into higher-return activities (Brown et al., 2006).

The next step sought to understand the temporal and spatial changes in land use and the underlying reasons for these changes. Here, participatory assessments were used to generate history lines for identifying and discussing crucial points of influence and change. The insights gained were then used to further develop and consolidate the preliminary classification from the *status quo* analysis performed in the first step, and to outline the overall strategic behaviour of the smallholder groups analyzed in the case studies. Subsequently, a complete linkage cluster analysis (with the WinSTAT statistics Add-In for Microsoft Excel, http://www.winstat.de) was performed to identify similarities between the outlined classes and strata and to determine the most appropriate variables for classifying smallholder behaviours (Vos et al., 2009). This process started with an a priori list developed through a literature review; the list included issues such as the combination and proportion of income portfolios, land-use components, livelihood assets, landholder motivations (Davies and Hossain, 1997; Ellis, 2000; Zoomers, 2001), the nature of the activities (Scoones, 1998; Swift, 1998; Ellis, 2000; Shackleton et al., 2008), the reproduction of livelihoods, the continuity of land uses, adaptation behaviour (Alwang and Siegel, 2003) and in- and outmigration (De Jong et al., 2006), especially rural-urban networking (Padoch et al., 2008).

Finally, the livelihood clusters that we developed were used to discuss future scenarios of land-use trajectories for the region, taking into consideration local responses to and influences on institutional and environmental settings compatible with local capacities and interests, particularly focusing on decisions about the use of trees and forests. This scenario analysis was applied as a multi-stakeholder dialogue oriented towards a set of key variables derived from the analysis of the case studies and the related institutional and environmental settings. The analysis also considered the more specific insights gained in the cases explored by individual PhD and MSc students. The scenarios revealed possible avenues for local decision making on land use, providing insights on the social and economic outcomes of such decisions and their likely consequences on interactions at the household and landscape levels.

Research on environmental aspects

The initial aim of the environmental research component of ForLive was to better understand the mutual linkages between smallholders and ecosystems (forests in particular) at the level of individual properties and landscapes. This approach was based on three assumptions derived from research published in the past few decades: (1) smallholders depend on the goods and services provided by nature, and trees and forests – particularly NTFPs – have a significant potential to contribute financially to local livelihoods (Shanley et al., 2008); (2) the loss and degradation of forest ecosystems – and thus the benefits that they provide – affect poor rural people more directly than urban populations (MA, 2005); and (3) smallholders contribute to the ongoing process of deforestation in the Amazon (Aldrich et al., 2006; D'Antona et al., 2006; Perz et al., 2006).

With the growing awareness of the dramatic increase in deforestation in the Amazon over the past three decades, the landscape dynamics and changes in the Amazon have become a central focus for environmental science and policy discussions, and more broadly for public opinion on a global level (Moran, 1993; Skole and Tucker, 1993). During this period, the major and unprecedented deforestation in the region (Rodrígues et al., 2009) has been associated with a wide range of positive effects such as economic growth, but also with negative effects on the socio-cultural integrity of poor forest dwellers, particularly indigenous groups (Wunder, 2001), as well as biodiversity loss (Tucker and Townshend, 2000; MA, 2005), soil degradation (Moran et al., 2000), and, more recently, global climate change (UNFCC, 2011) and the associated effects on the local poor (World Bank, 2008).

To tackle this mutual linkage between people and the environment, ForLive's environmental research component followed an inherently anthropocentric approach, considering humans as valuing agents that enable the translation of basic ecological structures and processes into value-laden entities (De Groot et al., 2002). In accordance with the Millennium Ecosystem Assessment (MA, 2005), these structures and processes include a wide range of ecosystem support functions (soil formation, nutrient cycling and primary production), natural resources (food, fibre, fuel, biochemicals and other biological products), regulation processes (pollution control, microclimate regulation, water cycle regulation and pollination) and cultural services (aesthetic, recreational, religious, spiritual and cultural heritage functions) that generate benefits for people at local and global scales in the form of security, basic materials for well-being, health, social relationships and freedom of choice of action.

To capture the different spatial scales of human-environmental interaction, the environmental research component was performed at both the household and landscape levels. At the household level, the research focused on the roles that different types of forests play in local livelihoods, particularly considering four types of resources provided by ecosystems to rural people (Cavendish, 1999): (1) consumption goods (e.g. wild fruits, wild vegetables, large wild animals, firewood, agricultural tools, pottery and wild medicines); (2) inputs (e.g. leaf litter, thatching grass and livestock browse and graze); (3) outputs (e.g. wild fruit sales, wild vegetable sales, construction wood sales and other wild good sales); and (4) durables and stocks (e.g. furniture, large utensils, firewood, construction wood and fencing). At the landscape level, the principal goal of the research was to describe actor-specific land-use strategies as a basis for better understanding the environmental effects and identifying thresholds to guarantee the maintenance of environmental services from human-affected landscapes (Godar et al., 2012b). In particular, this landscape analysis focused on a comparison between the two prevalent land-use models found in most of the case study areas: small-scale management of natural resources, often including diversified family agriculture practiced by smallholders living in forested areas or settlements, and the production of commodity goods such as cattle, timber, soybeans and oil palm by largeholders, including cattle ranchers, loggers and agroindustrial companies (Chomitz and Thomas, 2003; Margulis, 2003).

Both levels of empirical research explicitly acknowledged the fact that deforestation has converted a vast area of mature forest into a mosaic of agricultural lands, pasture and forests at different successional stages. It is estimated that approximately 20–50% of previously deforested areas in the Amazon are in different stages of succession (Lu et al., 2003). This succession is of crucial importance for the restoration of soil through the accumulation of biomass, the creation of organic matter in the soil and various positive soil-plant interactions (Moran et al., 2000); secondary forests also house many commercially interesting species for the production of timber and NTFPs (Smith et al., 1997). Therefore, our research investigated the identification and accurate description of the different types of degraded and secondary forests characterized by different degrees of human influence and the resulting consequences on the provision of goods and services.

Household-level research

Data for the household component of the study were systematically collected within the case study communities in Bolivia, Ecuador and Peru. We took into account the surrounding landscape matrix, given its ecological influences, particularly 'edge effects'. This research followed a work plan implemented in three steps (Robles, in preparation): (1) the application of a multidisciplinary landscape approach to determine landscape elements and locally relevant products sourced from trees and forests; (2) the realization of tree inventories in all identified forest types; and (3) the extrapolation of inventory results to the level of landscapes as a basis for the investigation of long-term tendencies and human-ecosystem interactions.

In the first step, semi-structured interviews and accompanied walks across smallholders' properties in the selected case study sites were used to start learning about local perceptions of the importance of various landscape elements in the provision of ecosystem goods and services and to identify locally relevant products derived from trees and forests. Next, the Multidisciplinary Landscape Approach (MLA) (Sheil et al., 2002) was systematically applied to these families to attribute more specific values to the environmental goods and services obtained from the different forest types. Local people were also asked about their perception of trends in environmental dynamics and thresholds for the provision of desired goods and services by the different forest types. Several different participatory tools such as the pebble distribution method were applied in this process. This sequence of exercises provided a temporal and spatial understanding of smallholders' perceptions of different tree- and forest-based elements of their production systems and surrounding landscapes.

On the basis of the results from the MLA analysis, forest inventories of the identified local forest types were performed, focusing on the species that were most important in terms of local use. These inventories were conducted in cooperation with the local land owner and supported by local experts on the identification of plant species found in the surrounding villages. The field samples were selected via local mapping exercises and interpretation of satellite images using the following criteria as landscape qualifiers: the predominance of woody vegetation, a high distribution within the study area and easy access. Age estimation by the farmers was used to determine a sequence of successional stages. To avoid 'edge effects' within the selected sample area, the inventory included forest stands that were located in larger patches of the selected mature and secondary forests. In practice, all the studied secondary forest patches were located in smallholders' agricultural fallows that had previously been cleared for subsistence agriculture. The average size of a sampled forest patch was approximately two ha, corresponding to the area typically managed by a family for the production of agricultural crops each year. In each case study, approximately 30 plots were established for each identified land-use element, with each plot representing a different successional stage. Canopy cover, strata identification and distances to major features were all recorded in the stand description. The geographical coordinates for each plot were determined with GPS. After an initial visit to each stand to verify its accessibility, appropriateness and relevance, a plot of 30×30 m was demarcated, ensuring that the distance from the plot to the edge of the stand was at least 30 m in any direction (Figure 1.3). In these plots, the measured trees and saplings were identified to the species level with the help of field assistants and available identification keys. Standard quantitative measurements (species, diameter at breast height (dbh), tree height and stem height) were taken. These measurements were used to calculate a set of parameters describing the structural characteristics of the stand.

In the final step, the data from the forest inventories were used to estimate the potential of different forest types for the provision of locally valuable goods and services at the landscape level. These estimates utilized available maps and

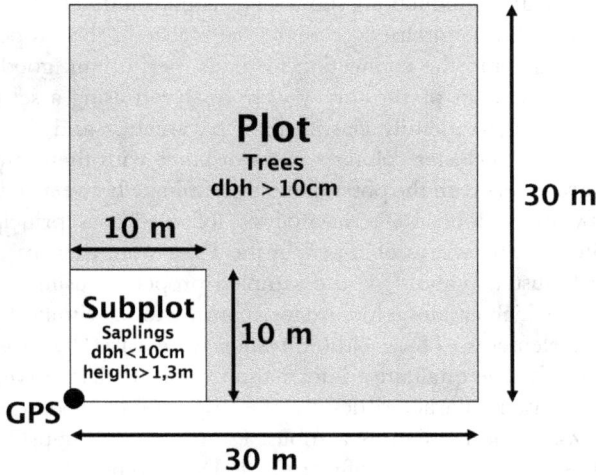

Figure 1.3 Sample design of the inventory plots at the property level (adapted from Robles, in preparation)

satellite images and information generated by participatory mapping exercises with the families involved in the case studies. Discrepancies that emerged between the interpretation of maps and satellite images and the results from the MLA exercise regarding the classification, size and location of the different forest types were used as an entry point to reflect on the possible reasons for divergent local perceptions and to revisit discussions of local expectations for the long-term development of human-ecosystem interactions. This process finally led to a modelling exercise in which the local families discussed their land-use options in relation to various assumptions about local trends in landscape dynamics resulting from possible institutional and environmental contexts in the future.

Landscape-level research

To assess the environmental outcomes of actor-specific land-use strategies at the landscape level, the tools and concepts of landscape ecology were applied in combination with socio-economic data from interviews with landowners and resource managers and the interpretation of satellite images. This component of the research followed three chronological working steps: (1) a classification of landowners; (2) the geospatial analysis of properties and landscapes considering different temporal and spatial scales; and (3) the calculation of landscape metrics.

In a first step, interviews were conducted to determine household characteristics and land-use strategies among the different actor groups (Godar et al., 2012b). The surveys aimed to include all actor groups relevant to the different study areas. Through personal communication with representatives of the identified actor groups, a representative sample of families and producers was defined by taking into account the predominant product (permanent crops, cattle, annual crops, timber or diversified production), the level of capitalization (subsistence-oriented, non-capitalized, low-capitalized or highly capitalized), the property size (small, medium or large) and the connection to roads (peri-urban, good connection or remote). The outcomes of the survey were analyzed using a set of multivariate statistical methods to identify classification parameters and discrete thresholds of property sizes to cluster colonists in accordance with their land-use patterns. First, the parameters with the potential to discriminate between actor groups were identified by an 'exploratory factor analysis' followed by a 'principal component analysis' (PCA). The scores obtained in the PCA were then used to perform a 'hierarchical cluster analysis' of the sampled properties using 'Ward's method' (Ward, 1963), which creates a hierarchy according to the similarity between pairs of observed elements. These cluster results were carefully cross-checked and consolidated with the qualitative information from the field surveys to aggregate quantitative cluster characteristics by the application of descriptive statistics, principally using the median as a robust estimator. The statistical analysis was performed using the statistical software R (R Development Core Team, 2005).

The geospatial analysis sought to measure the effects of the different actor groups on landscapes (Godar, 2009; Godar et al., 2012b) by attributing the properties of the sample areas to the defined actor clusters and overlaying their property boundaries on to land cover maps (Browder et al., 2008) obtained through remote sensing by integrating forest inventories and spectral data, as suggested by Vieira et al. (2003) and Lu (2005). Using available Landsat Thematic Mapper (TM) and Enhanced Thematic Mapper (ETM+) images (see Table 1.1), the landscapes were classified into the following categories: mature forest, advanced secondary forest, intermediate secondary forest, initial secondary forest, farmland, urban areas, roads, water and alluvial sediments. The images were atmospherically corrected using an improved, image-based *DOS model* (Chavez, 1996) and geometrically rectified using a 'second-order polynomial'. 'Histogram matching' (Du et al., 2002) was performed to transform the older images. Non-forest land uses were identified by unsupervized classification using the 'ISODATA algorithm' and ground truthing with a GPS device. The forest classes in the satellite images were discriminated by correlating vegetation structure data from forest inventories of 20×20 m plots with reflectance data extracted from 2×2 pixel windows at the same locations. For each plot, forest successional stages were discriminated using 'Canonical Discriminant Analysis' (Lu et al., 2003), resulting in numerical scores reflective of above-ground biomass and other parameters that increase with forest maturity. Multitemporal changes were mapped to facilitate understanding of long-term dynamics and spatially explicit degradation/restoration processes.

In a final step, landscape metrics were calculated as proxies for the ecological value of the actor-influenced landscapes, including environmental services such as biodiversity, gene flow, greenhouse emissions, the persistence and movement of plants and animals and energy fluxes related to hydrological services (Lindenmayer et al., 2002; Godar et al., 2012a). The metrics used were forest fragmentation (forest patch density and forest edge density) and forest core area based on anthropocentric influence (a distance of 300 m was used to estimate the area of

Table 1.1   Analyzed Landsat images from the selected study areas

	Medicilândia Brazil	Brasil Novo Brazil	Anapú Brazil	Pacajá Brazil	Riberalta Bolivia	Pucallpa Peru
Path/row	226/62 226/63	226/62 226,63	225/62 225/63	225/62 225/63	233/68	006/66
Timeline	1987, 1991, 1999, 2007	1987, 1991, 1999, 2007	1986, 1993, 1996, 2001, 2007	1986, 1993, 1996, 2001, 2007	1996, 2005	2006

potentially lightly degraded forests (Laurance et al., 2000) and a distance of 3,000 m was used to estimate the area of presumably inviolate core forests) and largest-patch indexes, as well as forest connectivity level and interspersion/juxtaposition indexes based on a distance between patches of 500 m (Laurance et al., 1997, 1998).

The methodology outlined above was applied to its full extent in the Brazilian case study area along the Transamazon highway (Godar, 2009, Godar et al., 2012a, 2012b). The intensity of research in the study areas in Peru and Bolivia was significantly lower and mainly served to challenge the findings obtained from the Brazilian study area. The study area in Ecuador was not considered in the landscape analysis, as the available satellite images suffered from excessive cloud cover, which hindered the proper interpretation of a sufficiently high proportion of pixels. Overall, more than 160 inventory plots were evaluated to discriminate forest types in the Landsat images, and a total of 27 multi-temporal maps were constructed for the five study areas, covering an area of approximately 50,000 km². These maps had an accuracy level of more than 82%, which is high compared with prior studies (Lu et al., 2003; Vieira et al., 2003; Lu, 2005).

Implementation of the research

This section outlines the principal strategic features of the ForLive study, explaining the chronology of research activities, the different levels of implementation and the organization of learning and reflection procedures, which constituted a fundamental prerequisite for the development of the insights presented in this book (Pokorny, 2003). In general terms, the ForLive project was carried out in three phases: (1) orientation; (2) research; (3) synthesis and dissemination (Figure 1.4). Each phase built upon the previous one.

During the orientation phase, we screened the study areas in Bolivia, Brazil, Ecuador and Peru to identify and describe tree- and forest-based socio-productive systems in which smallholders had managed to generate significant local benefits. The focus on 'promising' examples for local resource management originated from the initial expectation of elucidating the underlying reasons for these success stories as a basis for realistically assessing the potential for replication.

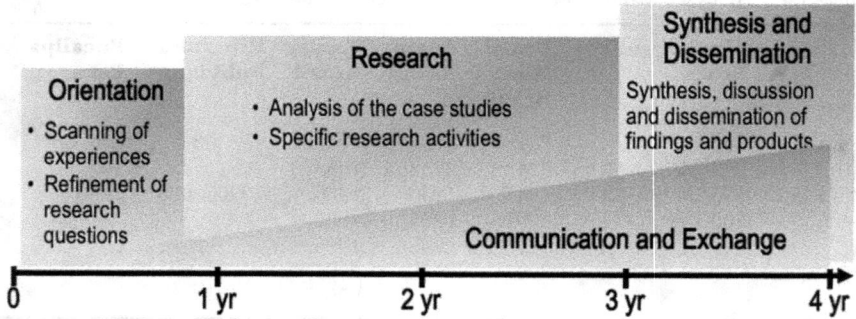

Figure 1.4   Structure of the main work plan for the project

In particular, we aimed to learn about the local capacities and institutional and environmental conditions necessary to achieve such ends. This phase included a systematic identification and review of available information about the study areas, the relevant actor groups (particularly the smallholders) and the principal challenges for local development. A sequence of workshops was organized in each study area with the participation of representatives from relevant local and regional actor groups and experts. These workshops had three principal goals: (1) establishing a basis for regional cooperation; (2) discussing and specifying relevant research questions; and (3) collecting suggestions for promising smallholder initiatives. Following the suggestions given by the workshop participants and individually interviewed policy makers, experts, technicians and local families, the project assistants then visited nearly 150 sites in the field and processed the information that they gathered in a common database. The project partners used the aggregated information from the workshops, interviews and field visits to decide on the specific research questions to be tackled by the different research components and to select 17 case studies for in-depth empirical analysis.

Lasting nearly three years, the research phase, which involved applying the analytical frameworks outlined above to conduct empirical research on institutions, local livelihoods and environmental issues, essentially constituted the project's core. In addition to the case studies and study areas, the research process included numerous additional experiments in the course of individual research projects. In this phase, the researchers continued to discuss the collected data and emerging findings with local collaborators. This process culminated in the synthesis and dissemination phase, in which the large body of gathered information was structured and synthesized. In addition, open research questions were identified and tackled by specific complementary research projects. Finally, the most important findings were published and, together with outcomes of more specific relevance, disseminated to smallholders, NGOs, universities, donors and policy makers in a sequence of local, national and regional events. In parallel, during the entire four years of the project, a continuous process of internal and external communication and exchange was undertaken to inform relevant partners in the study areas about ongoing activities and reflect upon and discuss preliminary findings. This mutual learning process gradually formed the basis for an effective synthesis of results in the final phase of the project.

Organization of the research

The research in ForLive concentrated on the analysis of a set of 17 in-depth case studies. These case studies served as a platform for four research processes implemented in parallel: (1) comparative research in accordance with the analytical framework outlined above; (2) research on specific topics using one in-depth case study at a time as an exploratory example; (3) technical research on topics of local relevance; and (4), a number of individual academic research activities on specific issues considering several of the in-depth case studies in combination with broader research questions. All four research processes were interlinked through a set of formal and informal mechanisms for communication and exchange to stimulate mutual reflection and learning processes across spatial scales (households, landscapes and countries), disciplines (institutions, livelihoods and environment) and actors (locals, technicians and researchers).

In-depth case studies

As a first step, the project made efforts to establish a basis for dialogue and mutual understanding with the families in each of the 17 sites selected for in-depth case studies. This was achieved by intense negotiation with the families about the objectives, contents, expectations and responsibilities related to the project, resulting in agreements of understanding. In addition, local and regional networks were developed and strengthened by systematically linking relevant people and organizations located in the case study areas. In each in-depth case study, we implemented the complete analytical framework outlined above. The data gathered here provided the empirical basis for the projects' efforts in comparative analysis and synthesis. One project assistant was assigned the responsibility of coordinating each case study and communicating and documenting the project's research activities, with the support of PhD students. The assistants were guided by field manuals developed for each research component. The information generated in each case was entered into a standardized cumulative database, the 'information package', which was made available to all researchers on the project's homepage.

Complementary research on selected topics with relevance for the in-depth case studies

In addition to the core research, all project assistants, mostly NGO technicians with a local university degree, were encouraged to develop their own individual research projects to be completed within the in-depth case study. This process was expected to stimulate the assistants' active involvement in the scientific dialogue and ensure their active contribution to the project's reflection and synthesis mechanisms. The assistants were intensively supervised by more experienced researchers from the project consortium. Generally, these project assistants' research projects were focused on the underlying reasons for the selection of the specific in-depth case study, which were used to develop a related research hypothesis. This hypothesis was consolidated by an extensive literature review for subsequent analysis on the basis of empirical data gathered from the case study. Finally, these studies also involved critical reflection on the relevance of the findings for the broader Amazonian context. In more general terms, the objective of this research component was to identify and critically reflect upon the paradigms concerning forests, smallholders and rural development. Thus, these research projects also contributed to the objectives of ForLive with specific empirical insights and considerations. The specific research topics are listed in Table 1.2. The findings of these research projects constitute the basis for a book project currently in progress.

Table 1.2   List of topics for complementary research on case studies

Case Study	Topic	Responsible researcher(s)
Bolivia		
12 de Octubre	Individual versus collective action as a small farmer forest production strategy in the *campesino* community of 12 de Octubre in the Bolivian Amazon region.	Huanger Avila
Buen Destino	The strength of community organization for the Management of Brazil nuts in the community of Buen Destino, in Northern Bolivia.	Claribel Quiette
Palmira	The adoption of agroforestry systems in the northern region of the Bolivian Amazon: a case study of small farmers from the community of Palmira.	Edgar Escalera Muchia and Teresa Oporto Daza
Buen Futuro	The role of the exchange of knowledge in the adaptation of colonist farmers to local conditions: a case study of a small colonist farmer in the Bolivian Amazon.	Vincent Vos, Claribel Quiette Quispe and Teresa Oporte Daza
Peru		
Curimaná (Amabosque)	The role of oil palm in forest use along the Neshuya-Curimana road axis in the Peruvian Amazon.	Carlos Sanchez
Caserío 7 de junio	The compatibility of the adoption of highly demanded commercial Amazon products with meeting food security requirements in small farmer production systems: the case of camu-camu (*Myrciaria dubia* HBK Mc Vaugh) in the community of Yarinacocha, Peru.	Erick Garcia
Carretera Federico Basadre (km 28.8)	Motivations for land-use change by small farmers in peri-urban areas of the Amazon.	Juan Pablo Ferreyros

Case Study	Topic	Responsible researcher(s)
Comunidad Nativa Callería	Evaluation of the logging agreements in three indigenous communities Callería, Nuevo Saposoa and Pátria Nueva.	Pío Santiago and Iván Icochea
Ecuador		
Pajanak	Factors affecting the success or failure in the establishment of forest plantations by small farmers in the Ecuadorian Amazon.	Marco Romero
Chinimbimi	Endogenous *silvopastoriles* systems as a source of income for small farmers in the Amazon.	Rosa Masaquiza
La Quinta Cooperativa (Luz de América)	Comparison between artisan and mechanized logging practices by small farmers in the Amazon.	Fredi Tandazo, Stefan Gatter and Marco Romero
Simón Bolívar	The potential for logging of pioneer species in secondary forests by small farmers in the Amazon: the case of pigüe (*Pollalesta discolor*) in the central-southern region of the Ecuadorian Amazon.	Juan Pablo Merino
Brazil		
Medicilândia	The role of the forest in rural development: the case of the smallholders in the Medicilândia Transamazon highway region.	Javier Godar
Majari	Informal logging agreements with small farmers: lessons for Community Forest Management.	Romy Sato and Max Steinbrenner
Cajarí	Constraints and opportunities for cooperatives in the processing and marketing of Brazil nuts from extractivist communities in the RESEX Cajarí, Amapá, Brazil.	James Johnson and Fabrício Nascimento
PAE Ecuador	The hands that touch the forest: the influence of forest management on internal community organization of extractive farmers from Xapuri, Brazil.	Patricia Cabalero and Fabrício Nascimento
Machadinho d'Oeste	Forest management agreements between loggers and communities: the case of Machadinho d'Oeste, Rondônia, Brazil.	Deryck Pantoja and Íran Veiga
Municipio de Cametá	The effects of the growing demand for extractive products on small farmers producing açaí (*Euterpe oleracea* Mart.) in the municipality of Cametá, Pará, Brazil.	Luciane Costas and Francisco de Assis Costa

Local research

In each of the in-depth case studies, the collaborating families, communities and, where relevant, accompanying governmental or non-governmental support organizations were invited to conduct supported research on topics of local relevance. The project facilitated the identification of local priority themes, the development and implementation of the research design and the analysis of data and compilation of findings through participatory action research processes. Intensive training was provided to the project staff to facilitate this complex process. In addition to a general workshop to explain the guiding principles and underlying concepts of this working approach, specific training workshops were organized in each of the five project areas to provide an orientation on participatory research approaches, including their application in the field. The implementation of this participatory research component was supervised and local researchers received on-site training from the experienced ForLive researchers.

In each in-depth case study, so-called 'Local Research Groups' were formed by the local families to develop and work on a commonly agreed-upon research agenda. The 'Local Research Groups' in ForLive focused mainly on technical questions such as the control of pests and diseases, the establishment or management of specific crops or trees and the control of fire; others were concerned with markets, particularly those for timber, or with community organization. Despite severe challenges in putting this approach into practice (Pokorny and Johnson, 2008c), this activity resulted in highly interesting outcomes at the local level. The process challenged researchers and smallholders to explain and justify their activities and thereby created space for mutual learning and discussions on perspectives, experiences and knowledge. Consequently, this research component not only built local research capacity and generated interesting information for the smallholders, but also enhanced the local relevance of the scientific research and promoted local families' access to the findings of the scientific studies. However, due to the technical character and process orientation of this research component, this book considers only a small portion of the information generated by the local research groups.

Individual academic research projects

In addition to the core case studies, there were also a number of individual academic studies, mostly implemented as research projects by PhD and MSc students from Europe and South America. Given the strong and immediate commitment of researchers in these initiatives, this component of the research was essential in achieving the project's objectives. The individual research projects essentially contributed to the comparative analysis of the projects' general research questions while also exploring additional, more specific research questions. Most interestingly, these individual studies went beyond the in-depth case studies and often considered a broader universe of experiences and additional research questions with relevance to the study areas, thereby significantly enriching the critical reflection on the working hypothesis. To effectively organize this individual research within the core research, a research matrix was developed that defined, for the array of research questions in these projects, the information required to answer these questions, the methods used to obtain the information and the researchers within the project consortium who were responsible for coordinating the analyses of the defined questions. In combination with the comprehensive database compiled from the in-depth case studies, this significantly broadened the empirical basis for analyzing both the research questions from the work packages and the individual research questions. Among all the individual research projects listed in Annex 1, this book is particularly focused on the findings of the nine studies presented in Table 1.3.

Table 1.3   Individual academic studies with high relevance for the synthesis phase of ForLive

Responsible Researcher(s)	Title of research	Methodological approach	Related Publications
Javier Godar	The environmental and human dimensions of frontier expansion in the Transamazon Highway	Comparative analysis of the economic, social and environmental effects of settlers and cattle ranchers on landscapes; GIS analysis of Landsat satellite images over three decades combined with field measurements of ~3,000 individual properties and interviews with 100 representatives of relevant actor groups in four municipalities along the Transamazon highway in Brazil.	Godar, 2009; Godar et al., 2008, 2012a, 2012b
Lisa Hoch	Do smallholders in the Amazon benefit from tree growing?	An analysis of the potential outcomes of tree growing for Amazonian smallholders; analysis of promising tree-growing initiatives with respect to strategies related to support, costs, benefits and challenges; interviews conducted with experts and local farmers and direct field measurements from the ForLive study regions and in Puerto Maldonado and Loreto (Peru), Pastaza (Ecuador), Santa Cruz (Bolivia) and the Brazilian states of Pará, Amazonas and Rondônia.	Hoch, 2009; Hoch et al., 2008, 2009, 2012); Pokorny et al., 2010a
Gabriel Medina	Moving from dependency to autonomy: an opportunity for local communities in the Amazon frontier to benefit from the use of their forests	This study explores the relationships between local traditional communities and external players in the Amazon frontier to analyze their influence on locals' ability to benefit from their forests to achieve local development and to develop autonomous (in contrast to dependent) options based on their own interests and capacities. These issues were assessed by analyzing six of the core case studies in Bolivia, Brazil and Peru, considering communities with and without strong relationships with external actors such as timber enterprises and NGOs.	Medina, 2008; Medina et al., 2008, 2009a, 2009b, 2009c

▶

Responsible Researcher(s)	Title of research	Methodological approach	Related Publications
Jes Weigelt	Reforming development trajectories? Institutional change of forest tenure in the Brazilian Amazon	Analysis of the scope of action of smallholders affected by agrarian reforms, focusing on the potential to benefit from their natural resources. The orienting principle is ontology, or the question of what I consider to be elements in the social world that possess causal powers, that can cause effects, particularly agents, social structures and discourses. The study considered empirical data from five case studies of agrarian reform in sites located in the frontier regions of Western Pará.	Weigelt, 2011
Marco Robles	Ecosystem goods and services from land-use systems of smallholders in the Amazon	To describe locally relevant goods and services from different tree- and forest-based production systems and analyze the long-term influence of different management options by analyzing eight of the core case studies in Ecuador, Peru and Bolivia.	Robles, in preparation
Oscar Llanque and Vincent Vos	Forest-related livelihood strategies of smallholders in the Amazon	Evaluation of the importance of forests for local people and the potential of forests to contribute to local livelihoods; application of the sustainable livelihoods framework adapted to more specific research questions about the role of forests in all core case studies selected in ForLive given their promising nature and the presence of neighbouring families and communities who have not received external support.	Vos et al., 2009; Peralta et al., 2009
Gabriel Medina and Benno Pokorny	Financial analysis of 'Community Forestry'	Evaluation of the financial viability of 'Community Forestry' initiatives for sustainable timber management as implemented in pilot projects; analysis of the costs of timber management based on models developed in accordance with primary empirical data from the case studies, interviews and secondary data from the eight most promising experiments in the PPG-7 subprogram ProManejo in the Brazilian Amazon.	Medina and Pokorny, 2008, 2011; Pokorny et al., 2012

Responsible Researcher(s)	Title of research	Methodological approach	Related Publications
Cesar Sabogal	The legal and institutional framework for 'Community Forestry'	A comparative survey of official regulations regarding forest management, land ownership and rural development impacting smallholder forest management conducted in Bolivia, Brazil, Ecuador and Peru to (1) identify and compare the legal frameworks that generate impacts on small-farmer forest management, (2) evaluate the implications of the legal frameworks with respect to the promotion or restriction of forest management activities by small farmers, and (3) make recommendations to improve or harmonize the legal frameworks for small-farmer forest management.	Sabogal et al., 2008a; Carvalheiro et al., 2008; Ibarra et al., 2008; Martínez Montaño, 2008
Cesar Sabogal, Wil de Jong, Benno Pokorny and Bas Louman	'Community Forestry' in tropical America: Experiences, lessons learnt and perspectives	A group of outstanding researchers and professionals working on 'Community Forestry' in Latin America joined forces to gain a better understanding of what is meant by 'Community Forestry', who its protagonists are, the critical dimensions of the enabling environment that influences 'Community Forestry' and the challenges that families face when pursuing 'Community Forestry'. Based on the multiple experiments and the profound knowledge of the authors, this book project reviews the outcomes of this effort and assesses the achievements and challenges of 'Community Forestry' in Latin America.	Sabogal et al., 2008b; Pokorny and Johnson, 2008b; Pokorny et al., 2009, 2010b

Reflection and exchange

In light of the complex research agenda, the large number of researchers involved, the interdisciplinary nature of the team and the intention to actively involve all relevant actor groups – particularly the smallholder groups – the exchange of and critical reflection on the insights, ideas and experiences of all involved were fundamental to guaranteeing an effective learning process. The project aimed explicitly to mobilize tacit knowledge from the project partners and equip people with the empirical insights gained from the case studies as well as their own valuable experiences (Polanyi, 1983; Flick, 2008). This process included regular field visits to local families in the case study areas and the organization of regular multi-stakeholder meetings for reflection and discussion (Figure 1.5).

Internal project workshops provided the project partners in all study areas with the opportunity to communicate and coordinate their research agendas and to present and discuss ongoing research. These workshops were also used to evaluate overall project progress. In total, there were three internal project workshops. At the beginning of the project, the coordinators of the research components and partner organizations met in Ecuador for the *Initiation Workshop* to adapt the project outline approved by the European Commission to the specific local contexts and the demands and interests of the partners. After eight months, the entire project team met again in Bolivia to define the guiding research questions, harmonize the agendas of the research components and select the case studies on the basis of a regional screening process for promising smallholder initiatives. Before and after the harmonization meeting, the project assistants and researchers linked to the different project regions were trained for field research. In the third year, a *Project Retreat* was held in Peru to prepare for the synthesis phase by discussing preliminary findings from the different research initiatives. A work plan to fill in missing information was defined.

PhD workshops brought together doctoral students and supervisors to discuss the progress, problems and agenda of the ongoing research. Larger workshops were held at the beginning of the second year and during the third year of the project. The first workshop primarily addressed the research outlines and possibilities for collaboration; the second workshop was largely dedicated to the analysis of preliminary results and the definition of strategies for the dissemination of the findings.

A series of local seminars and thematic workshops were also held in each partner country to present and discuss the research outcomes within local networks including farmers, policy makers, universities and NGOs. In the first project year, these local events aimed to establish the local networks; the events held in the second and third years mainly discussed the findings of ForLive and the resulting policy implications.

Finally, in the fourth year, in the majority of the European and South American partner countries, seminars and conferences presented the findings of ForLive and disseminated recommendations to the target groups. In December 2008, the

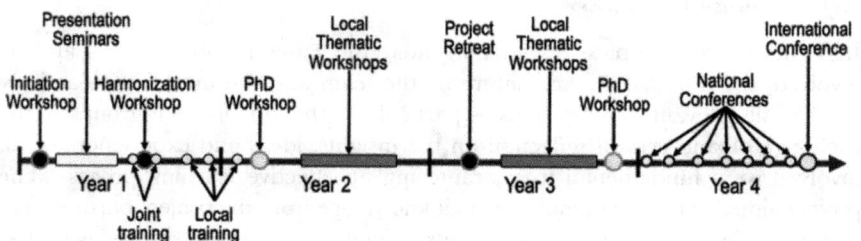

Figure 1.5   Principal project events organized by ForLive

project culminated in an international conference on the role of smallholders in forestry development in the Amazon, which was held in Belém, Brazil.

Consolidation of the key findings

As outlined above, regular opportunities for shared reflection with ForLive researchers, the families involved in the in-depth case studies and representatives of the relevant actor groups were essential for the challenging and consolidation of research findings emerging from the field research. In addition, in the final year, the project also invested in a systematic assessment of the key findings by adapting the methodological approach of pattern-matching (Yin, 2009). As described by Terluin (2003), the pattern matching consists of a sequence of three steps: (1) the specification of a hypothesis as a predicted pattern of events; (2) the collection of information on these events in case studies; and finally (3) the comparison of the observed events with the predictions of the hypothesis. In addition, the methodological approach foresees the definition of premises for and conditions under which the predicted events are expected to occur.

As a basis for systematic assessment, the main preliminary conclusions from the comparative analysis of the case study findings and the more timely insights gained in the individual research projects were discussed during a project retreat in Peru at the beginning of the third year of the project. These discussions resulted in a simplified model of landscape dynamics in the Amazon as a basis for the definition of a set of working hypotheses regarding the main lessons learnt. The initial idea of the assessment was to challenge these working hypotheses in all study areas and assess the scope for extrapolation and generalization of the model and the need for differentiation and local adaptation. Thus, it was expected that the various study areas would yield empirical evidence and knowledge to consolidate, adapt or refute the hypotheses. In total, 15 working hypotheses were defined in the form of cause-and-effect relationships, with a special focus on the pattern of events. For each hypothesis, we discussed when and under which prerequisite conditions the predicted events might occur. Finally, a set of indicators for empirical assessment in each of the study areas was defined (Table 1.4).

Table 1.4   Verification of the key messages (adapted from Blum, 2009)

#	Working hypothesis	Context	Indicators
1	Smallholders' productive activities create, within their individual properties, a diverse mosaic of land-use components, including temporarily preserved forest patches.	Forested landscapes; spatial dominance of smallholders;	Mosaic of small land-use patches; forests within smallholders' properties.

▶

#	Working hypothesis	Context	Indicators
2	Within smallholders' production systems, secondary forests play a crucial role in maintaining soil fertility, in subsistence and, in the case of accessible markets, in the generation of complementary income, particularly from non-timber forest products. Local agroforestry systems are highly efficient in using available resources.	Sufficiently large properties; absence of capitalized actors.	Existence of secondary forests and trees; active management of trees; local markets for products from secondary forests and trees; ongoing production within the properties.
3	Smallholders respond spontaneously to emerging opportunities for benefits (markets, credit or partnerships) by adapting their production systems. This response implies, however, that they are open to changing even successful systems when available options change. Production components are exchangeable, including forest management, which is only considered as long as it is sufficiently attractive or supported.	Fluctuations of market prices; change in market demands; adaptation to farm production systems.	Abandoned plantations; smallholders seeking direct (short-term) benefits; market prices for current farm products higher than current prices for past products.
4	The increased presence of logging companies accelerates the degradation of smallholders' forests, independently of the quality of the regulatory framework.	Access roads; traditional contexts.	Number of timber companies; degree of law enforcement; location of the forests; deforestation rates.
5	Smallholders receive low prices for the timber that they sell to traders and logging companies, even when harvesting is based on an authorized management plan. In the negotiation process, companies generally take advantage of their competitive advantages in the marketplace. Nevertheless, the majority of smallholders highly welcome the possibility of generating several income streams.	Presence of logging companies; scarcity of legally accessible forests.	Share of costs and benefits; type and number of management plans.
6	If smallholders use their forests in accordance with their traditional management schemes, they do not exceed environmental thresholds, even when they commercialize their products on local markets.	Limited presence of timber companies; forested landscapes; long-term settlement.	Volume of forest products; prices on local markets; quality of forests on smallholders' properties.

#	Working hypothesis	Context	Indicators
7	Current policies systematically favour capitalized actors but typically ignore the needs and capacities of smallholders. Thus, the vast majority of smallholders do not have a realistic chance of success in these new economic environments or of maintaining their socio-productive systems.	Accessible areas; economically relevant recourses available.	Public services, in particular credit programmes; infrastructure investments; requirements for accessing public benefits; private investments; tendencies of land tenure.
8	Political and economic elites are systematically using their power to appropriate land and resources in frontier areas. Consequently, the proportion of land and resources under the control of capitalized actors is constantly increasing, and large-scale land-use activities including cattle, soybean and biomass farming dominate, greatly accelerating environmental degradation.	Accessible areas; economically relevant recourses available.	Economic interests of local politicians in power; personal relationships between economic and political elites; land-use changes; land tenure situation.
9	Smallholders suffer from unfavourable conditions that hinder their ability to improve their situation. In addition to a notorious lack of capital, they often suffer from insufficient internal communication and are poorly organized. They are often embedded in paternalistic relationships with more powerful actors. Their access to public services, especially credit and proper education, is limited.	None.	Development index; level of social organization and institutional representativeness; power relationships; education system; credit programmes.
10	The direct influence of smallholder-targeted development initiatives, often limited to a few families, is marginal within the general dynamic in frontier areas, which is dominated by the private sector, which in turn is generally not sufficiently controlled by the state.	Existence of development initiatives.	Investments in smallholders; number and characteristics of development initiatives; number and characteristics of beneficiaries.
11	Development projects show promising initial results, but after the project ends and external support leaves, the innovations are abandoned by smallholders.	Existence of development initiatives.	Goals of development initiatives; duration and dynamic of development initiatives.
12	Externally initiated development projects tend to replace rather than to improve the local socio-organizational systems and thereby contribute to a loss of endogenous practices, knowledge and mechanisms. This situation is particularly true with regard to tree- and forest-related initiatives.	Existence of tree- and forest-related initiatives.	Goals of development initiatives; compatibility with local systems. Degree of dependence on external support; diversity of management practices.

▶

#	Working hypothesis	Context	Indicators
13	Within development initiatives, the supporting organizations tend to establish paternalistic relationships with smallholders, who have little chance of developing their own positions regarding development in accordance with their own interests and capacities.	Existence of development initiatives.	Modes of interaction between external organizations and smallholders; degree of compatibility between local needs and initiatives; development discourses employed by the actor groups.
14	Under specific conditions, smallholders show their capacity to develop and fight for their own ideas.	Strong external pressure on resources in traditional forest contexts.	Level of organization; history of social movements; presence in media and public policies.
15	Well-organized smallholders can establish alliances with influential actor groups to effectively articulate their interests in public policies; however, in the course of this alliance they adapt to the interests and discourses of their allies.	Social organization, alliances with external actors.	History of social movements; presence in media and public policies; discourses; hierarchies.

To challenge the 15 hypotheses, a three-month survey was conducted in Bolivia, Peru, Ecuador and Brazil, in which 125 governmental officials, extension agents, local and international scientists, representatives of smallholder organizations and smallholders were interviewed. The insights gained from these interviews were complemented by information from locally available literature and field observations. Contradictory information was discussed in depth with key informants. The final interpretations of this triangulation process were then used to draw conclusions about each working hypothesis in internal project meetings and public seminars held in each study area.

This procedure for assessing key hypotheses contributed in two ways to the synthesis phase of the project: first, by generating a basis for the careful verification of possibilities for the extrapolation and generalization of research findings; and second, by providing additional empirical information on the project's research questions, especially those related to the broader institutional contexts of the study areas. Therefore, the results and questions emerging from this assessment process served as an important input for the final sequence of reflection events, where the lessons learnt were discussed with representatives of all relevant actor groups.

Study sample

In consideration of the project's limitations with regard to resources and time, our research focused from the beginning on regions characterized by a relatively high abundance of smallholder families and a relatively long duration of settlement to ensure a wide range of experiments and effective sampling. Moreover, for logistical reasons, study areas in proximity to the offices of local partners were preferred. The following five study areas were defined: the northwest region of Pando in Bolivia, the northeastern region of Pará in Brazil, the state of Acre in the Western Amazon, the region near Macas in Ecuador and the Pucallpa region in Peru (see Plate 2).

In each study area, we intended to select a number of in-depth case studies representative of promising efforts by smallholders to effectively use trees or forest resources for local development. Due to a lack of conceptual clarity and information, it was not possible to define the selection criteria a priori. Therefore, in an iterative process during the orientation phase, the project partners compiled the incoming empirical data from the review of experiments within the study areas and then decided on the final set of criteria for study selection. Then, during the *Harmonization Workshop* at the end of the first year, the local project partners, in accordance with local expertise and institutional interests, selected a total of 17 in-depth case studies forming the core of the project's comparative research.

In view of the small number of case studies and the subjectivity of the selection process, the selected in-depth case studies cannot be considered to be a fully representative sample reflecting the diversity within the region. Nevertheless, they do reflect a meaningful part of the wide range of existing experiments and contexts within which smallholders live. This sample was complemented by several other experiments that were considered in the course of individual academic studies. To generate an understanding of this broad sample, which generated the insights presented in this book, the following sections describe the relevant features of the study areas, the selected in-depth case studies and the specific experiments considered in the individual research projects.

The five study areas

The five study areas of ForLive represent typical realities for highly dynamic agricultural frontiers in the Amazon. In these areas, smallholders interact – to highly varying degrees – with other actors representing a wide range of lifestyles, capacities and interests. These other actors, denoted in this book by the term 'external actors', include large-scale agroindustrial, forestry, mining and energy businesses, migrants, politicians, traders, development organizations, schools, health agents and others. The majority of the analyzed experiments were located in the type of rural contexts that were generally characterized by a relatively low level of infrastructure. However, it is important to note that the project did not collaborate with families in extremely remote and isolated areas, where the influence of external actors is still considered to be very limited.

Bolivia

The study region in Bolivia was located near the city of Riberalta, in the northern states of Beni and Pando. Over the past decade, this study region has become more connected with the remainder of the country and the greater region. The initiation of the construction of the 'Northern Corridor', a macroscale construction project creating a major road between countries, has promoted the connection of this region with Brazil and Peru. However, as the realization of these plans is slow, access during the rainy season remains difficult. Consequently, for a large number of these communities, the river continues to be the principal mode of transportation (see Plate 3).

A huge proportion of this region remains covered by vast areas of primary forest. Since the decline of rubber tapping, the harvesting of Brazil nuts (*Bertholletia excelsa*) has become the mainstay of the regional economy, forming the principal source of income for many families and communities. However, cattle ranching has begun to gain prominence in the local economy, in addition to the exploitation of lumber and rapidly growing agricultural production. In particular, the production of soybeans, coming from neighbouring regions in Brazil, has started to expand.

The population is predominantly indigenous, with antecedents whose families came to work as labourers during the rubber boom. During this period, the region was settled by migrants, mainly from Beni and Santa Cruz, but also from Europe, Brazil and Peru. More recently, migrants from the Andean region have moved to Riberalta, attracted by the growing trade here (and their hard lives in their mountainous home regions). The families generally cultivate their properties using slash-and-burn techniques and, particularly in families of indigenous origin, harvest products from commonly owned forests for commercialization in local markets. In several cases, such as with Brazil nuts, the products are also exported. Increasingly, the land-use dynamic is influenced by cattle ranchers and agroindustry and timber companies. In many districts, powerful families of 'rubber barons' controlled huge tracts of land ('latifundios') by establishing systems of semi-slavery of families (called 'baraqueiros') for the production of rubber, and this system continued with the harvesting of Brazil nuts. Since the 1920s, many of these powerful families have managed to maintain their influence and continue exploiting the region's resources, despite the fact that they formally lost their ownership as a result of several land reforms (Assies, 1997).

Brazil

The two study areas in Brazil were relatively far away from each other. Although both areas represented 'typical' frontier contexts, each region demonstrated rather specific land-use dynamics. The areas included extractive reserves (RESEX) and neighbouring areas with informal land tenure regimes, agro-extractive settlement projects (PAE) and colonization projects.

In an extractive reserve, the government holds the property rights to the land but recognizes the rights of traditional communities to access the land. Thus, local families are allowed to use the forests on the basis of authorized management plans. In Porto de Moz, in Pará, and in the study area of Amapá, local families harvested NTFPs such as Brazil nuts, rubber and timber from collectively owned primary forests. The majority of these families practised shifting cultivation, and several engaged in small-scale cattle ranching, often in smaller plots for individual use by each family. Extensive large-scale cattle ranchers and loggers exerted pressure on the land and forest resources. In Acre, the production areas, including the forests, were used individually for the production of nuts and rubber. Cattle also played an important role. In this area, and in the study area of Pará, there was still a large portion of primary forest remaining.

In more recently settled regions, often in government-directed settlements, the majority of primary forests have been converted for other land uses, with families engaged in diversified production systems including annual crops, perennials and cattle ranching in individually controlled plots of 50–100 ha. This was the case for the study area located in the municipality of Medicilândia along the Transamazon Highway, in Pará, where families began to settle in the 1970s over the course of several colonization projects. In this study area, the smallholders produced annual and perennial products within individually owned lots, with cocoa as the main commercial product. Capitalized actors concentrated on cattle ranching. Here, conflicts over land and resources were commonplace and have recently been exacerbated by the growth of agroindustry (soybean, rice, oil palm and sugarcane farming) and the Belo Monte hydroelectric dam project.

Ecuador

In Ecuador, the study areas were located in the provinces of Morona Santiago and Pastaza, near the city of Macas in the south-central region of the Ecuadorian Amazon. Unlike the other study areas, the Ecuadorian study area was located at approximately 1,000 meters above sea level and possessed fertile deep volcanic soils. The absence of a distinct dry season here means that the farmers rarely used fire as a production technique. Although the infrastructure was better than in the other study areas, many communities still did not have access to reliable roads or electricity.

These areas have traditionally been occupied by the 'Shuar' ethnic group, who typically cultivate individual plots and commonly manage forests for timber and NTFPs. Since the 1950s, these areas have been settled by migrants, mainly from upland regions to the south, as part of government-supported colonization programmes. The majority of farmers were engaged in logging, cattle farming and the cultivation of a variety of agricultural crops. Migration to other countries (mainly the US and Spain) for foreign income was also a rather common phenomenon in the region and constituted an important component of the local dynamic. Consequently, many families in the region benefited from overseas remittances and off-farm income, but they also suffered from a shortage of labour force that greatly limited agricultural productivity.

Because of greater recent investment in the construction of roads, especially in the north and central regions of the Ecuadorian Amazon, the area is becoming more attractive for the timber and mining industries and, increasingly, for the production of palm oil on large plantations. However, oil palm exploitation has had a limited impact in the study region compared with other parts of Ecuador.

Peru

In Peru, the study region was located close to the city of Pucallpa, in the Province of Ucayali. This area was characterized by three distinct contexts: first, relatively intensive peri-urban agriculture; second, settlements along the roads dominated by Andean families producing agricultural crops and cattle for national markets; and third, seasonally flooded forest lands along the Ucayali river populated by indigenous communities ('shipibo konibo') practising shifting cultivation, fishing and the harvesting of forest products. In more remote areas, the cultivation of coca (*Erythroxylum coca*) was still a factor. Logging companies were engaged predominantly in illegal timber trading with these communities. Depending on how close each location was to transport routes, the colonists along the roads had considerable access to public services, whereas the indigenous communities living in remote areas were still relatively isolated.

In areas adjacent to the roads, palm oil cultivation has become one of the principal economic activities. Petroleum and gas exploration have also contributed significantly to the development of the regional economy. Since 2006, this dynamic has been strongly supported by policies of the Peruvian government that were systematically designed to attract foreign investors focusing on export markets.

In-depth case studies

Within the study areas in Bolivia, Brazil, Ecuador and Peru, more than 150 promising smallholder initiative sites were visited; a large proportion (44%) were located in Brazil, approximately one-fifth (22%) were in Bolivia and Peru, and 12% were in Ecuador. More than half of the initiatives (57%) were located within areas still characterized by continuous forest cover, whereas the remaining initiatives were located in landscapes characterized by somewhat fragmented and sometimes almost completely transformed landscapes. There was an emphasis on experiments with settlers (52%), but the review also included initiatives by indigenous groups (except in Brazil, where research in indigenous areas was rendered impossible by a highly complex and bureaucratic formal permission process) and traditional communities including riparian and extractivist groups. Several groups had small farms with forest areas of less than five ha, but the majority of holdings were between 50 and 100 ha, whereas several other, mainly indigenous, groups had access to collectively owned forests of up to several thousand hectares. Half of the groups used their forests based on collective rights, but the other half had individual rights over resources. The vast majority of families were organized into associations, cooperatives or unions; only 19% were not. More than 80% of the initiatives received external support (approximately a quarter received intensive support) from government or non-governmental organizations, mostly for sustainable timber management in natural forests, but also for plantations and agroforestry systems. However, the sample also included families working without external support and experiments related to the production of NTFPs such as fruits and fibres.

The 17 in-depth case studies selected from this range of screened initiatives showed great heterogeneity in their social and environmental features (Table 1.5). The sample included different tenure and land-use regimes, including both communally managed and individually managed production systems in protected areas, recent frontiers, older and younger settlements, different types of small-holders (indigenous groups, traditional families and small farmers), and social groups and individuals with only small plots as well as those with large tracts of forested land. Additionally, the selected case studies showed different levels of external support: very intensive and long-term, intensive but temporary and no significant support.

Table 1.5 Main characteristics of the 17 in-depth case studies

Location	Smallholder category	General description	Main productive activities	External support
Bolivia				
12 de Octubre	Traditional agro-extractivist community with indigenous roots ('campesinos'(1))	3,600-ha common area of forest (formal request on 16,378 ha) with individual farm plots.	Harvesting of Brazil nuts, logging, shifting cultivation and cupuazu (*Theobroma grandiflorum*) in agroforestry systems, livestock.	Intensive NGO support for agroforestry.
Buen Destino	Indigenous community (Cavineña)	Common forest land (7,000 ha) in indigenous territory (468,117 ha).	Harvesting of Brazil nuts, shifting cultivation, small-scale cattle ranching.	Moderate NGO support for forest management.
Buen Futuro	Campesinos of indigenous origin and one migrant family	Individual land (200 ha) in a mosaic of primary and secondary forest and fields.	Shifting cultivation for rice, cassava and maize, agroforestry systems, small-scale livestock, harvesting of Brazil nuts.	Intensive NGO support for agroforestry.
Palmira	Campesino community	Individual land (50 ha) with primary forest remnants.	Shifting cultivation, agroforestry systems with cupuazu and fruit trees, small-scale livestock ranching, harvesting of Brazil nuts.	Intensive long-term NGO support for agroforestry.

▶

Location	Smallholder category	General description	Main productive activities	External support
Brazil				
RESEX Cajari in Amapá	Traditional community	Extractive reserve of approximately 500,000 ha of forest with individual user rights.	Harvesting of Brazil nuts and palm hearts, shifting cultivation, fruit trees in home gardens.	Extensive NGO and government support.
Majari, near RESEX Verde para Sempre in Pará	Traditional riverside families ('riberinhos')	9,100 ha of commonly owned forests with individual plots of 50–100 ha without legal tenure.	Shifting cultivation, fishing, use of forest products.	No external support.
Medicilândia, on the Transamazon highway in Pará	Settlers within government colonization programmes	Individually owned plots of 50–100 ha with highly fragmented forest patches.	Agroforestry systems with cocoa, cattle ranching, shifting cultivation.	Extensive government support for agriculture and cocoa.
PAE Ecuador	Ex-rubber tappers in a legalized settlement	Total area of 7,000 ha with individually owned plots of 500 ha.	Harvesting of Brazil nuts and rubber, cattle ranching, shifting cultivation.	Intensive government and NGO support for timber and forest management.
Ecuador				
El Eden	Indigenous families mixed with some settlers	Commonly owned land divided into individual plots of approximately 50 ha.	Slash and mulch agriculture, hunting, logging, small balsa-wood plantations (*Ochroma pyramidale*), agriculture.	Extensive NGO support for timber management and plantations.
ACAPP	Settlers	Individual properties (up to 50 ha) with small patches of primary forests (20%) but mostly secondary forests.	Harvesting of pioneers such as pigüe (*Pollalesta karstenii*) in secondary forests, fish farming, fruit production, e.g. naranjilla (*Solanum quitoense*).	Extensive NGO support for plantations and timber management.
Chinimbimi	Settlers	Individual land of 30–100 ha with degraded secondary forests.	Palm fibre stands and cattle.	No support.
Wachmas	Indigenous community (Shuar)	Commonly owned forest area with individual user rights for lots of approximately 45 ha.	Subsistence agriculture and timber harvesting.	NGO support for timber management.

Location	Smallholder category	General description	Main productive activities	External support
La Quinta Cooperativa	Settlers	Individual land with secondary and fragmented primary forests.	Commercial agriculture (cassava, corn, naranjilla, etc), timber harvesting, cattle.	NGO support for timber management.
Peru				
Callería	Indigenous community (Shipibo Konibo)	Commonly owned primary forest (3,650 ha).	Shifting cultivation, fishing, timber harvesting (FSC-certified),	Intensive NGO support for timber management.
Campo Verde	Peri-urban settlers	Individual land (approximately 100 ha), pasture and highly degraded forests.	Cattle, extraction of palm fruits, e.g. buriti (*Mauritia flexuosa*), bee keeping, fish farming,	Extensive NGO support for NTFPs.
Pueblo Libre	Settlers	Individual land plots (approximately 37 ha) with primary forest remnants and secondary forests.	Oil palm plantations, timber, honey, cattle,	NGO support for agroforestry.
7 de Junio	Settlers	Individual land plots (20 ha) with NTFP plantations.	Production of camu-camu (*Myrciaria dubia*) and other fruits,	NGO support for plantations.

(1) In Bolivia, the term 'campesino' is used generically for small producers with indigenous roots who over time lost their original cultural identity, as indicated by the fact that these families typically do not consider themselves to be indigenous. Another frequently used term for these people is 'mestizo', denoting mixed European and indigenous descent.

Additional case studies

As outlined in Table 1.3, a number of individual academic research projects were conducted within the framework of ForLive; these projects contributed to the empirical basis and comparative analytical potential of the project by addressing case studies and study areas in addition to the core sample presented above. The studies by Javier Godar (2009), Lisa Hoch (2009), Jes Weigelt (2011) and Gabriel Medina and Benno Pokorny (2008, 2011) in particular generated data that contributed substantially to the findings presented herein.

Godar (Godar, 2009; Godar et al., 2012a, 2012b) studied the economic, social and environmental effects of settlers and cattle ranchers on landscapes by analyzing Landsat images over three decades and combining these data with field measurements of 3,000 individual properties and interviews with 100 representatives of relevant actors in four municipalities along the Transamazon highway in

Brazil, each of which was characterized by different proportions of specific actor groups (Table 1.6).

Table 1.6   Main characteristics of the municipalities considered in studies of socio-environmental effects of colonists on landscapes (adapted from Godar, 2009)

	Medicilândia	**Brasil Novo**	**Anapú**	**Pacajá**
Predominant actor group (occupied area)	Small-scale settlers (71%).	Medium-scale settlers (48%).	Large landholders (43%), small-scale settlers (46%).	Large landholders (41%).
Principal product	Agriculture, principally cocoa.	Cattle ranching (cattle hot spot in the Amazon).	Cattle ranching, agriculture.	Cattle ranching.
Institutional characteristics	Settlements restricted to 100-ha plots; well-organized farmers.	Settlements in properties up to 500 ha, dominance of settlers from the South.	Settlement programmes for properties of up to 3,000 ha, significant conflicts.	Settlement programmes for properties up to 3,000 ha, credit programmes for cattle, conflicts.

Hoch (Hoch, 2009; Hoch et al., 2009, 2012; Pokorny et al., 2010) analyzed the potential for tree growing by Amazonian smallholders, investigating more than 100 tree-planting experiments in the project's study regions as well as in Puerto Maldonado and Loreto (Peru), Pastaza (Ecuador), Santa Cruz (Bolivia) and the Brazilian states of Pará, Amazonas and Rondônia. She conducted interviews with experts and local farmers and undertook field measurements to determine the smallholders' strategies with regard to support, costs, benefits and challenges (Table 1.7).

Table 1.7   Main characteristics of the sample considered in the study of smallholders' tree-growing initiatives (adapted from Hoch, 2009)

112 interviews with experts and technicians (N)

Researchers (28); NGO workers (27); government employees (25); international experts (9); company managers (9); smallholder representatives (8); others (6)

Field measurements of 90 tree-growing initiatives in 80 case studies (N)

Population: Traditional and indigenous communities (31); recent settlers (59)
Access: By river (19); By road (5); both (67)
Type of support: Preliminary and broad (83); continuous, more intensive (7)
Source of support: NGOs (50); government: (37); companies (3)
Year of establishment: 1980s (25); 1990s (34); After 2000 (31)

Weigelt (2011) analyzsed five agrarian reform initiatives in Western Pará aimed at securing smallholders' rights to land and forests to determine the effects on local institutions. The results from interviews and field observations within the case study areas were complemented by a review of secondary sources (Table 1.8).

Table 1.8 Main characteristics of the areas considered in the study of local effects of reform initiatives (adapted from Weigelt, 2011)

Name	Ademir Federicce	Renascer	Riozinho do Anfrísio	Virola-Jatobá	Verde para Sempre
Type	Sustainable Development Project	Extractive Reserve	Extractive Reserve	Sustainable Development Project	Extractive Reserve
Size (ha)		211,741	736,350	24,000	1,290,000
No. of families in 2005		600	50	200	2,500
Year of creation	Not created	2009	2004	2003/2005	2004
Municipality	Medicilândia	Prainha	Altamira	Anapú	Porto de Moz
Poverty rate in 2000 (%)	48.58	78.55	37.66	64.60	68.57

Medina and Pokorny (2008, 2011) analyzed the costs of timber management based on models developed in accordance with primary empirical data generated from interviews and secondary information from eight of the most successful 'Community Forestry' initiatives in the Brazilian Amazon (Table 1.9).

Moreover, this book explicitly takes into account a comparative survey of the legal and institutional frameworks impacting smallholder forest management; this survey was conducted in each of the four countries under the coordination of César Sabogal from the Centre for International Forestry Research (CIFOR) (Sabogal et al., 2008a; Carvalheiro et al., 2008; Ibarra et al., 2008; Martínez Montaño, 2008, Pokorny et al., in press a) and included work by Benneker (2008) on 67 community forest enterprises in Bolivia.

The insights and empirical evidence gained from all these studies contributed in two ways to the research and synthesis process of the project: first, by enriching discussions with new ideas and hypotheses, which were then challenged in the core case studies, and second, by providing additional opportunities to reflect upon the main findings of the core research in a broader context, thereby strengthening the basis for interpretation and conclusions.

Considerations regarding the challenges of conducting collaborative research

ForLive followed an innovative approach of inter- and transdisciplinary research based on the analysis of case studies while simultaneously attempting to fulfil the requirements of disciplinary academic science, with the conventional outcomes (such as publications) demanded by funding bodies. This was a complex and challenging process, both from the methodological point of view and from the individual perspectives of the researchers involved in the project (Pokorny and Johnson, 2008c). The project demanded that the researchers take on tasks outside their own area of particular expertise and become involved in direct exchange

Table 1.9 Main characteristics of the pilot initiatives considered in studies of the costs and benefits of 'Community Forestry' (adapted from Medina and Pokorny, 2008)

	Oficinas Caboclas	Boa Vista d. Ramos	Mamirauá	Pedro Peixoto	Ambé	Costa Marques	Alto Acre	Porto Dias
Location	Pini near Santarém, Pará	Amazonas	Tefé, Amazonas	Acrelândia, Acre	Santarém, Pará	Rondônia	Xapuri, Acre	Acre
Population	Small traditional community, ex-rubber tappers		Several traditional communities living in remote areas	Governmental settlement project with individual plots of 80 ha	Cooperative in National Park	Cooperative in National Park	Ex-rubber tappers (RESEX)	Ex-rubber tappers in individual properties (PAE)
Equipment	Coll. Dead wood; Animal skid.	Chain saw	Chain saw; Mobile saw mill	Mobile saw mill; Chain saw	Heavy machinery		Machines subcontr.	Machines subcontr.
Skidding	100% local	100% local	100% local; 100% local	100% local; 100% local	Machines subcontr.	All work subcontr.	Machines subcontr.	Machines subcontr.
Area (ha/yr)	1; 40	40	22; 10	4; 4	1000	500	450	100
Volume (m³/yr)	5.52; 5.52	83.4	67.34; 24.13	11.68; 11.68	9,109	4,380	2,700	650
Final product per yr	240 pieces of furniture	40 m³ boards	30.25 m³ boards; 10.84 m³ boards	5.84 m³ boards; 4.91 m³ boards	9,109 m³ logs	4,380 m³ logs	1,116.62 m³ logs	268.82 m³ logs

with the smallholders and the relevant organizations involved in the case studies. This approach resulted in a significantly higher level of engagement and greater intellectual challenge for the researchers, but also caused them to have less time available for their own specific research agendas. In light of this situation, several researchers showed resistance to actively including non-academics in the research process and correspondingly in appreciating non-academics' views and capacities in their own work. In practice, for many of the senior researchers, travel to the study regions was limited to participation in project meetings and conferences. In this sense, the project confirmed the often-criticized separation between 'academia' and practice and revealed a great (enormous in some respects) distance between science and poor peoples' realities that is widened by an elitist and theoretical academic system (Pokorny and Johnson, 2008a, 2008c).

Several of the local researchers – often academically qualified people working as technicians at NGOs – were strongly tied to their traditional roles as members of a broader strategy, acting as project implementers for international funding organizations. These technicians experienced significant difficulties in using the space provided by the project for reflection and critical discussion on prevailing beliefs and opinions; instead, they often expected clearly defined working plans from the project coordinators. The observation of these difficulties emerging in the research process confirmed that research is essentially a social process. It became obvious that all project participants were shaped by strong formative impressions in their educational and professional experiences related to the classical research and development approaches, and these impressions tended to govern personal and professional behaviours. Obviously, the courses offered at universities or other technical training schools generally place more emphasis on theory and, most importantly, academic education and the corresponding work provide few opportunities to engage in contact with the rural poor. This situation hinders effective communication between academics and local people, who are essential as the basis for relevant and effective pro-poor research and the dissemination of innovations. The researchers' experiences in ForLive demonstrated that changing this situation is not easy, as the majority of researchers feel comfortable with the 'status quo', and neither research institutions nor granting agencies provide incentives for change.

Nevertheless, for many of the people involved in ForLive, the project managed to initiate a process of self-reflection; as a result, the majority of participating researchers began to critically reflect on their own positions, preconceptions and roles, leading them to more openly consider local views and realities. It is extremely important that the direct collaboration with smallholders contributed to an increased awareness of the distance between 'academia' and the potential beneficiaries of such research. Thus, the researchers' experience in taking responsibility for the project and in effectively collaborating with the rural poor reflects both the general difficulties imposed by the social system as a whole and the possibility of overcoming these barriers.

The principal methodological challenge of the project was to meaningfully bring together information at different empirical and theoretical levels sourced

from a wide range of highly diverse study settings and to avoid illegitimate generalizations. In practice, this process required the combination of findings from the
analysis of, for example, an indigenous community living in a remote forest area in
Bolivia with findings from a peri-urban settler family close to the city of Pucallpa,
Peru, with 200,000 inhabitants. To this end, the project developed a standardized
framework for comparative analysis of the 17 in-depth case studies representing a
(much) broader gamut of similar situations. We also established a range of mechanisms for communication, exchange and reflection among the ForLive researchers
acting at various levels in different contexts and for systematic discussion with the
representatives of relevant stakeholder groups in all study areas. In this respect,
knowledge management was found to be extremely challenging. The attempt
to establish a shared database compiling the relevant information from the case
studies and study areas for the common use of all researchers was only partly
successful despite significant input. In fact, the database was useful in guiding
the field researchers and documenting the gathered data but was not effectively
used for the research process itself. It became obvious that all researchers, without
exception, strongly preferred to gather the data relevant for their research
questions on their own, even when specific information was previously available.
Obviously, individual interests and competencies strongly dominated the research
agendas of the project partners.

Scientific exchange and cooperation, however, became reality during the
events organized for discussion and reflection. These facilitated exchanges highly
motivated the participants to discuss and interpret empirical findings from other
researchers and to embrace lessons learnt in their own research work. Thus, it was
possible to effectively feed tacit knowledge and competencies into the interpretation and synthesis process of the main project and into the individual research
projects of the different working packages. The quality of these mutual learning
processes, however, depended significantly on the availability of professional
individuals with the ability to communicate and assimilate valuable contributions,
as well as enormous efforts in the coordination of the reflection and synthesis
processes. Of note, some researchers involved in the research process did not
display this capacity, and consequently hindered the development of even more
fruitful collaboration for the benefit of the project.

Ultimately, the many and enthusiastic contributions from the internal and
external project partners in the case studies and reflection events guaranteed
the interpretation of the empirical information at the case study level. Thus, the
insights and interpretations presented in this book resulted from a transparent
process of scientific discussion and rely on a large body of empirical information
and profound theoretical considerations. Readers interested in further exploring
the profound analytical basis and a more specific characterization of the study
areas, case studies and other studied experiments are referred to the numerous
scientific products published as PhD, MSc or BSc theses and in scientific journals
(see Table 1.3 and the list of references).

2 Smallholders' actions and activities

This chapter highlights the huge diversity of practices and strategies employed by Amazonian smallholders to utilize their natural resources, particularly trees and forests. In general, ForLive confirmed that, in rural settings, trees, forests and their products – especially non-timber forest products (NTFPs) – play an important role for many families, fulfilling subsistence needs and providing income through local commercialization (Belcher and Schreckenberg, 2007). In particular, for the poorest segments of Amazonian forest societies, forest-related income functions as a safety net when crops have failed or as a source of cash in the case of emergencies such as illness or for the payment of school fees (Angelsen and Wunder, 2003; Neumann and Hirsch, 2000). However, although almost all the families in the case studies worked with trees in some manner or another, both inside and outside the forest, the majority of smallholders were more involved in agriculture and livestock than in the exploitation of forest products, with the exception of several indigenous groups and extractivist communities. Consequently, many families tended to define themselves more as farmers than as forest managers. Thus, the environmental benefits provided by forests, such as pollination of cultivated crops, protection of the soil and water supplies and restoration of soil fertility in swidden fallows, were often more important for these families than the forest products (Ghazoul and Sheil, 2010). Regarding the financial effect of forests on family income, the studies found a rather limited potential, especially when the family had a small property and only timber was considered. Overall, the smallholders used low-input management practices that were effectively adapted to local realities and available capacities. It became evident that the families involved in the case studies tended to opportunistically take advantage of any emerging opportunities for commercialization and support, and they adapted their land use accordingly.

Diversity of livelihood strategies

The case studies reflected the wide range of livelihood strategies adopted by smallholders in the rural Amazon (Vos et al., 2009). The findings of the ForLive study suggest that, in spite of using similar arrays of land-use components, each family followed its own specific strategy with a complex combination of different production activities (see Plate 4). These strategies typically included measures for generating revenue and subsistence production, and they were fulfilled by the various members of a family.

We identified 22 principal activities that we classified in the Amazon small-holders' livelihood portfolios (Table 2.1). This classification does not include further differentiation between specific products such as the large variety of NFTPs, nor does it include activities of minor importance, those performed only once or twice a year, or those that require only minimal time input such as the collection of medicinal plants. We classified the principal production activities into four major categories: (1) agriculture systems based on modern technology, typically requiring new technology and high upfront investment; (2) agroforestry systems including traditional slash-and-burn techniques, silvopastoral systems and tree plantations; (3) extraction systems to remove and commercialize forest products; and (4) river-utilization systems, principally for fishing. Within all these categories and activities, we found examples of both commercial and subsistence uses, generally occurring in conjunction with each other. Additionally, families often combined their production activities with product processing and refinement activities, most frequently with forest products. Off-farm activities also played an important role; these included allowances paid by other small-holders and external employment with large landowners and logging companies in addition to government payments and support by family members working in urban areas or abroad.

In all the analyzed cases, families simultaneously engaged in a number of activities so that their lifestyles constituted diverse 'portfolios'. Most families performed between 6 and 13 of the 22 principal activities. Food production was by far the most important activity for all families, especially when considering the large investment in manual labour. In fact, in almost every case, the producers cultivated food (90%) and/or crops for forage (75%). Approximately half of the families owned livestock or bred small animals, but forest products were also used for subsistence and for the generation of income. More than 80% of the families collected firewood or NTFPs for commercialization or hunted. Timber extraction activities were performed by 42% of the sampled families.

The wide range of activities practised by families, in combination with the specificity of individual livelihood portfolios, clearly makes it difficult to achieve a meaningful classification. Many previous attempts at classification have taken into account characteristics such as livelihood strategies (Chomitz et al., 2006; Fearnside, 2008), production systems (Browder et al., 2004), property size (Margulis, 2003), land tenure dynamics (Browder et al., 2008), strategies for intensification, extension or migration (Scoones, 1998; Swift, 1998), and the degree of dependence on natural resources (Ellis, 2000). However, in accordance with our findings, these attempts suffered from the exclusive consideration of only one or two specific aspects and therefore necessarily failed to grasp the complex reality of Amazonian smallholders. For example, a producer cultivating fibre-producing plants to sell to a broom manufacturer while also producing food in shifting culti-vation, keeping several cows and engaging in other productive activities could be

Table 2.1 Portfolio of activities and products identified in the ForLive case studies

Activities	Principal products
Agriculture	
Low investment	Corn, rice, beans, vegetables, meat
High investment	Soy, rice
Animal breeding	Meat, eggs
Livestock/ranching	Meat, milk
Agroforestry	
Slash and burn	Food crops (cassava, rice, corn)
Home gardens	Fruit trees, poultry, vegetables
Agro-silviculture	Cultivation of perennial plants for commercial purposes (cocoa, coffee, etc)
Apiculture	Honey, wax
Silvopastoral	Meat, milk, fibres, wood, Brazil nuts
Tree plantations	NTFPs, palm oil, timber, firewood
Forest extraction	
Wood cutting	Firewood, timber
Collection and harvesting	Firewood, honey, medicinal plants, nuts, fibres, fruits, etc
Cocoa	Cocoa
River products	
Fishing	Fish
Aquaculture	Fish, crab, shrimp, etc
Processing	
Artisanship, handicrafts	Clothing, household items, paintings, art, etc
Refinement	Food, buildings, furniture
Off-farm labour for external parties	
Unskilled labour for neighbours	Daily wages/per diems
Unskilled labour for employers	Per diems
Employment in logging companies	Salary/wages
Employment in other sectors (public services, help, security)	Salary
Pensions, remittances from migrant relatives	Money

considered an NTFP collector, an agriculturist, a rancher, a commercial producer or a subsistence farmer.

In view of the insights gained in the in-depth case studies, Oscar Llanque, one of the ForLive researchers, developed a classification approach aiming to capture more of the complexity of the individualized and diverse livelihood strategies of Amazonian smallholders. He proposes a set of five combinable criteria: (1) the main purpose of production; (2) the degree of specialization; (3) the proportion of off-farm income; (4) the willingness and capacity to adopt innovations; and (5) the degree of cultural adaptation to environmental conditions, which is strongly related to the duration of residence in the region.

With regard to the *main purpose of production*, we distinguished two general strategies: production for commercialization and production for subsistence. In the case studies, we found a wide range of intermediate situations between these two extremes; however, the majority of families displayed a clear emphasis on a single purpose. In their attempt to benefit from market opportunities arising with the advance of the frontier, families showed a clear trend of gradual replacement of subsistence strategies with marketing strategies. This dynamic towards increasing commercialization was particularly prominent in families who had arrived more recently to the region. Nearly all such families were intensively engaged in the marketing and accumulation of goods, whereas traditional families who, by definition, had spent significantly more time in the region, relied more often on subsistence and survival strategies. That said, an interesting phenomenon was observed with the more recently settled families. After their arrival in the region, these families started to integrate local practices and especially strengthened the subsistence components in their portfolio of production activities. In parallel, they managed to adapt their production systems to the specific local environmental conditions surprisingly quickly and effectively. For example, in forested areas, this adaptation process often resulted in the systematic use of several forest products. Despite the limited number of samples observed and considering the huge diversity of individual situations and the cultural heterogeneity of families in the Amazon, this observation is indicative of the existing capacity of smallholders to adapt to new environmental, social and economic contexts.

The smallholders involved in the in-depth case studies also demonstrated different *degrees of specialization*. We noted a clear trend in that, over time, all producers increasingly concentrated their efforts on a smaller array of activities and/or started to produce for specific markets (Figure 2.1). Nonetheless, even at this level of specialization, individual families still produced a wide number of different products and services such as papaya, herbs, nuts, NTFPs, wood, timber, tourism, handicrafts and so on (Peralta et al., 2009). This higher degree of specialization in terms of allocated workload and income generation was associated with the continuously increasing involvement of families in commercial markets. Thus, specialization can be understood as a direct response to growing market opportunities. However, this process was accompanied by a much slower trend away from the original low-input, low-risk systems towards production systems requiring higher capital input, often financed through credit. Despite this process of specialization, nearly all families continued to produce food for their own consumption and to extract forest products when possible, initially as a strategy to save money for necessary purchases but also increasingly to reduce market risks (Ferreyros and Medina, in preparation). In general, although smallholders were commercializing and accordingly specializing to a greater degree, they were also maintaining a diversity of production components including agriculture, livestock, fisheries and extraction of forest products to meet both subsistence and market needs. This maintenance of diverse production activities was found to be an important and nearly universal feature of smallholders' livelihood strategies.

Another trend observed in all study areas was a significant increase in the proportion of smallholders' livelihoods coming from *revenues generated outside* the

Proportion of time spent on the primary activity (% of available family working power)

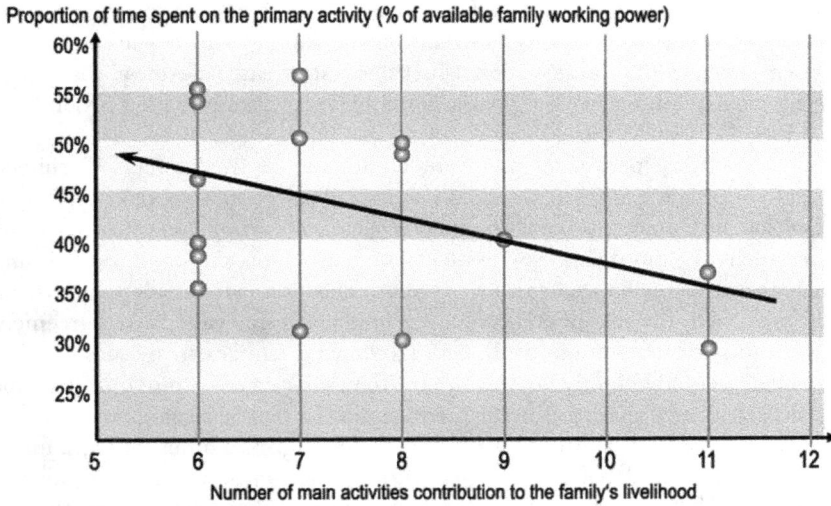

Figure 2.1 Proportion of time spent on the primary production activity in relation to the total number of activities in the family livelihood portfolio

smallholders' own properties. In many study areas, the payment of old-age pensions was a particularly important source of cash income, as also observed by Dethier et al. (2010). Although the underlying public pension schemes suffered from significant administrative deficits and high fiscal costs, many older people benefited from such schemes, which often became an important pillar of the family's cash income. Moreover, albeit to a minor degree, governments supported smallholder families with direct payments such as the 'Bolsa Familia' and 'Fome Zero' programmes in Brazil, through which approximately 12 million families currently receive funds for school scholarships in exchange for the guarantee that their children attend school and are vaccinated (see, for example, http://www.mds.gov.br/bolsafamilia). Peru also enacted policies designed to address poverty and assist the poorest among the approximately 50% of the population that lives below the poverty line. For example, the FONCODES (Fondo de Cooperación para el Desarrollo Social) programme was intended to generate local capacity to improve livelihoods through investment in facilities and infrastructure, and the support of microbusinesses and small-scale enterprises. In Bolivia, due to wide-reaching political reforms, the government has mostly focused on attributing a significant portion of its national budget to municipalities, each of which is tasked with addressing pressing internal issues. This funding originated from taxes imposed on hydrocarbon exploitation and from funds that were established to offset debts with international lenders. An important source of rural cash income channelled through these programmes originated from the National Emergency Employment Plan, which was also extensively implemented in Bolivia's tropical lowland prefectures and municipalities (Fuentes et al., 2005). In addition to government programmes, off-farm employment with large landowners or logging companies has also played a significant role. Many families also depended on financial transfers from relatives working in the rapidly growing urban centres (Padoch et al., 2008), large national metropolitan areas, or even abroad, with international remittances being most common in the study region around the Ecuadorian city of Macas.

Another critical parameter in the differentiation of smallholders' livelihood strategies was their *willingness to adopt technical and organizational innovations*. In fact, as a result of ambitious road construction projects, increased contact with external actors such as companies, governments and development organizations and better access to media and means of communication, smallholders have been confronted with massive organizational-technological innovations. In agreement with the observations made by Rogers (2003), the families in the case studies responded very differently to the opportunities emerging from this dynamic (e.g. access to new markets, credit programmes or technical assistance) due to their varying cultural and economic realities. As a gross simplification, families showed three typical behavioural patterns: (1) high opportunism, i.e. families who were ready to overhaul and replace their traditional production systems with new technologies and business structures; (2) careful openness, i.e. families who selectively adopted innovations to improve their own system and more effectively satisfy their own demands but generally avoided drastic changes; and (3) refusal, i.e. families who greatly hesitated to change their production system and lifestyle, either as a conscious rejection of large-scale changes or due to a lack of the skills and resources necessary for major change, thereby avoiding the risks associated with such changes.

As the last classification criterion, we suggest considering the degree of *cultural adaptation to environmental conditions*. In the majority of the case studies, the smallholders showed an awareness of the potential and limitations of their natural resources, including issues such as soil fertility, the growth potential of their crops, the range of useable forest products, etc. In fact, all families showed a considerable level of tacit traditional knowledge as a basis for decision making with regards to land use. Nevertheless, the study clearly demonstrated that the degree of adaptation to the socio-environmental context was strongly related to the *length of time that the smallholders had resided in the region*. Thus, the descendants of indigenous groups that settled the region in pre-Columbian times and those from families who arrived in the region during colonization by the Conquistadors mostly worked with production systems that were highly compatible with the environmental conditions, a phenomenon that was further pronounced by the fact that these actor groups often lived on and utilized relatively large, still-forested areas. However, families in more recent settlements also demonstrated the capacity to adapt to and change from environment-degrading production practices – often established in response to public incentives – to more sustainable practices better adapted to the specific socio-environmental conditions. However, this potential for adaptation, although it existed from the first day of each family's attempt at social reproduction and improved well-being, often became manifest only in the long term, after the original forests had already been transformed for other land uses.

In general, the families in the study demonstrated a lack of awareness about natural resource processes that facilitated sustainable use, as the smallholders' perspective was more focused (and increasingly so) on the commercialization of agricultural and forestry products than on the conservation and sustainable use of natural resources. The smallholders knew little about the regeneration of trees and the importance of genetics, and we also perceived a certain level of 'naïveté' regarding, for example, the recuperation potential of the forest and the long-term effects of the improper use of resources. Rarely did we find families who systematically invested in the search for opportunities to make their land use more compatible with environmental conditions to more effectively exploit the natural production potential over the long term. Thus, many of the families in the case studies worked with pre-established methods of production without sufficient awareness or consideration of specific environmental conditions and natural thresholds. This observation was particularly true with respect to the production of cattle but also relevant to the harvesting of forest products.

For some families in the study areas, particularly those who migrated into more remote areas, the degradation of their – often not formally owned – properties was even an intrinsic component of their livelihood strategy. After converting the standing natural forests in the areas they occupied for other land uses, most often cattle ranching, they sold the cleared land and moved on to new frontiers in search of new unoccupied land in still-forested areas. Often, these migrating families were systematically supported by more capitalized actors for the purpose of illegal land-holding (see Chapter 4). Nevertheless, the vast majority of families in the case study areas, although they endured extremely difficult conditions characterized by low soil fertility, difficult market access and an often complete lack of financial and technical support, managed to establish meaningful land-use systems on their properties, with effective, long-term use of available natural resources in accordance with their specific socio-economic conditions.

Due to the focus of the project on so-called promising cases, it is not possible to define in quantitative terms the proportions of smallholders belonging to the different classes outlined above. In addition, given the great diversity of situations and actors, we caution that the outlined criteria only express trends and thus, as with all other attempts at a meaningful classification of smallholders, they brush aside the actual complexity of the realities of these groups. Nevertheless, using these criteria to classify the quite heterogeneous sample examined in the project, it appears possible to define nine categories of principal livelihood strategies typical of smallholders in the Amazon (Figure 2.2).

Diversity of local practices in the use of land, trees and forests

The conversations held with families in and around the case study sites about the role and importance of trees and forests revealed that smallholders do not necessarily differentiate between agricultural land and forests. The regular alternation between agricultural cultivation and forest fallows as an intrinsic part of shifting cultivation agriculture by definition already integrates forest and agricultural

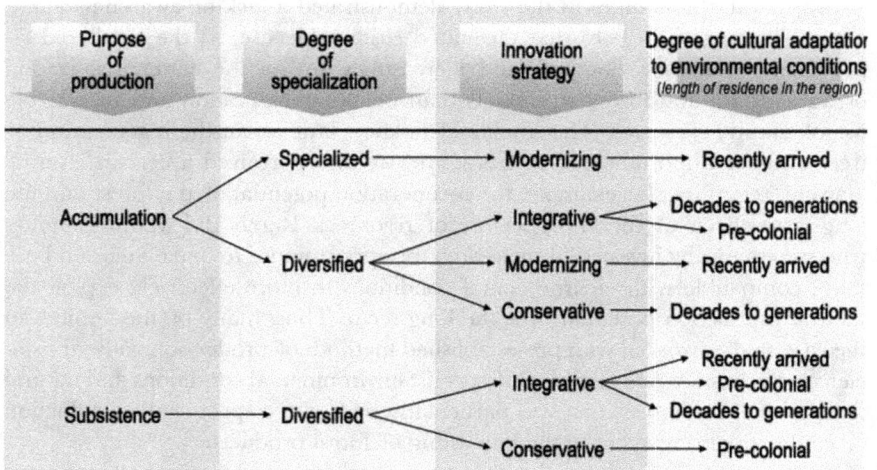

Figure 2.2   The nine principle livelihood strategies of smallholders in the Amazon as determined by purpose of production, degree of specialization, innovation strategy and duration of residence in the region

uses over time. In many of the case studies, families also developed production systems in which the cultivation of agricultural crops and the management of trees occurred side by side. For example, several families maintained fruit trees in their agricultural fields, established highly diversified home gardens, and worked with silvopastoral systems. In nearly all cases, the smallholders considered the existence or regeneration of trees as being valuable for them in their agricultural practices. Consequently, they avoided burning in areas with a high density of valuable trees and they actively protected and promoted the growth of trees with subsistence uses in their agricultural fields. The families often did this not only in the expectation of certain products of commercial – or, more frequently, subsistence – value, but also because of the trees' environmental functions, such as the provision of shade and/or soil and water protection. In fact, in all the case studies, the families valued the goods and services that trees and forests provided and thus included trees in their management and cultivation of agricultural crops. The care of trees was almost intrinsic to these families' activities, as they recognized its importance for their social, economic and environmental value. None of the families performed a systematic inventory or kept regular records of their tree resources, but the large majority knew their forests well and were often able to name the location of every single tree of interest. They also showed knowledge and a willingness to respect the ecology of certain tree species, although only for those in which they had an immediate personal interest.

A comparison between smallholders who received external support for the management of trees and forests and those who did not receive such support revealed that the smallholders' approach to working with trees deviated significantly from the suggestions made by NGOs and governmental agencies (Montero,

2007, Hoch et al., 2009, 2012). For example, with regard to plantations, the producers selected for the case studies generally favoured growing single trees, often following opportunities created by naturally emerging regeneration. Consistent with smallholders' understanding of trees as a complement to and support for the agricultural production component of their land management systems, in the majority of cases trees were found throughout properties, including in the agricultural fields. In contrast, nearly all external organizations contacted in the study areas favoured the systematic planting of seedlings in accordance with standardized technical guidelines, in separated, often larger, areas. In practice, the external organizations often separated forests and trees from the land.

The findings of the case studies confirm that Amazonian smallholders possess knowledge about forests, forest products and the management and cultivation of trees outside the forest. Although the smallholders often lack knowledge of the long-term ecological, social and political nuances of land-use decisions, in practice, these families have developed and continue to develop functional systems characterized by a huge array of technologies and practices to obtain benefits from the land, trees and forests. In almost all of the areas visited during the project, producers not only exploited timber and NTFPs, but also consciously took advantage of the environmental services provided by trees and forests to benefit their agricultural production.

Our field research indicates that each family develops its own combination of an impressive variety of land-use practices, which are adapted to the specific socio-environmental context. The extent of families' adaptation to forest management practices ranged from the adoption of simple techniques and harvesting practices to the implementation of more sophisticated silvicultural strategies or complex management rules aimed at achieving long-term goals. As described above, the locally employed practices that we observed in the case studies (for examples, see Box 2.1 and Plate 5) were generally characterized by significant integration with other land-use components, high compatibility with the specific situation of the family, a strong preference for low-input strategies and, as a direct consequence of this preference, a high degree of flexibility. The possible scope and potential of such local management strategies can be ascertained by studying the example of a cupuazu-centred agroforestry system near the Riberalta in Bolivia, where up to 21 tree species were found in farmers' small plots, with a density of up to 400 trees per hectare; one plot featured as many as 40 high-value Brazil nut trees (*Bertholletia excelsa*) (Robles, in preparation).

Box 2.1 *Examples of local practices for the use of trees and forests (Hoch et al, 2008, 2009; Vos et al, 2009)*

Stimulating the growth of valuable tree species by the selective cutting of lianas (frequently observed)
   Protecting individual Brazil nut trees (*Bertholletia excelsa*) by maintaining naturally regenerated trees in agricultural fields, avoiding fire damage from

slash-and-burn agriculture, and refraining from cultivating areas with high tree densities ('cavineños' indigenous group in the Buen Destino community, Bolivia)

Individually protecting the regeneration of fibre palm (*Aphandra natalia*) in silvopastoral areas, with the aim of gradually increasing the abundance of palms for fibre production (producers in Chinimbimi, Ecuador)

Continually refining techniques for the harvesting of Joshin pocote tree bark, which is used to prepare painted crafts (native community in Callería, Peru)

Developing a system for the production of cocoa (*Theobroma cacao*) that allows for natural regeneration and enrichment, planting of up to 200 trees per hectare of valuable timber species such as mahogany *(Swietenia macrophylla)*, ipê (*Tabebuia spp.*), bagassa (*Bagassa guianensis*) and cedar (*Cedrela odorata*) to provide shade and serve as a reserve for future generations (producer in Medicilândia, Brazil)

Developing silvicultural techniques to fight imperata grass (*Imperata sp.*), a weed that grows in degraded areas, hindering agriculture, increasing the risk of fire and hampering the regeneration of forest species (producers in Northern Bolivia)

Promoting the regeneration in pasture areas of timber trees such as freijó (*Cordia sp.*) and ipê (*Tabebuia sp.*) to provide shade for livestock and as a future income opportunity (producers in the Ecuadorian Amazon and along the Transamazon highway)

Transplanting and natural regeneration of fruit trees into home gardens for family consumption: mango (*Mangifera indica*), jackfruit (*Artocarpus heterophyllus*), açaí (*Euterpe oleracea*), uxi (*Endopleura uchi*) and cupuazu (*Theobroma grandiflorum*)

Intensively cultivating umarí (*Poraqueiba sericea*) for emerging fruit markets by planting, seeding and fostering natural regeneration (producers throughout the Peruvian Amazon)

Enriching natural stands and enhancing production for emerging external markets by seeding, planting and fostering natural regeneration of açaí (*Euterpe oleracea*) in Brazil and camu-camu (*Myrciaria dubia*) and the burití palm (*Mauritia flexuosa*) in Peru

Moreover, in the case studies specifically focusing on forest use, the harvesting of timber and NTFPs occurred spontaneously depending on the specific situation in time and space. Generally, the harvest was selective and involved minimal mechanization. To keep the work input low, the smallholders consciously considered the specific characteristics of the harvesting area. The accessibility of an area played an important role when making decisions on strategies for forest use. Generally, areas nearer to a road or a river suitable for transport were managed with a more intensive harvesting scheme. Often, forest products were left unharvested when they were too far away, prices were too low or simply because they were not the priority crop.

In the case studies involving traditional communities or indigenous groups collectively managing their forests, there were always locally developed norms and regulations for the use and sharing of resources. However, these norms were typically not formalized and they were often limited to specific aspects of use and management. These local regulations responded to the local conditions and took into account both the ecological characteristics of the individual species and the ecosystem itself, as well as the socio-economic conditions at the site. In the case of indigenous groups in particular, spirituality towards the forests was often an important consideration. The tree species locally considered most important were protected by more specific regulations. Many local norms also carried a strong social significance because they were often intended to ensure that all families had access to the resources and that there was an equitable distribution of costs and benefits. Consequently, these local regulations showed a distinct long-term perspective.

Still, the analysis was very clear in showing that producers rarely took advantage of the full productive potential of trees and forests. Although some producers have created and adopted important technical knowledge, in general, there was a certain reluctance by the smallholders to make use of the technical knowledge provided by experts, especially investing in even simple silvicultural techniques to increase the productive output of the trees. Thus, despite their functionality and local relevance, the majority of forest production systems employed by the producers also presented a great potential for optimization through easily implemented, low-cost techniques.

Addressing the limited potential of forests for the generation of household income

The central idea of the concept of sustainable forest management for smallholders, also called community forest management, is that the rational use of forests (especially timber) can generate income that is attractive enough that producers value the resource, thereby creating an incentive for smallholders to conserve the resource as an economic foundation of their livelihood (UNCED, 1992; Sabogal et al., 2008b). However, the ForLive case studies demonstrate that, despite the importance of the forest for local livelihoods, in particular for the poorest families living in more remote regions, examples in which families survive mainly from the use of forest products were rare. Apparently, the forest has potential for generating additional income but rarely provides the main foundation for local livelihoods. This was particularly true for many settler families whom we visited in the project: often they transformed forests without harvesting any trees for commercial purposes. Another indication of the comparatively low income potential of forests for smallholders is the fact that the price of deforested land in the study areas was generally significantly higher than the price of forest land (Hecht, 2005; Carrero and Fearnside, 2011), although other reasons such as user restrictions, accessibility and tenure situation also play a role in the forests' use and value.

The research conducted in ForLive demonstrates, in agreement with the findings of Vedeld et al. (2004) (see also at the Poverty Environment Network – PEN: http://www.cifor.cgiar.org/pen/), that the forest's contribution to family income was proportionally larger for (financially) poor families who, to a large extent, lived in more remote areas. In several case studies, the forest contributed substantially to the family income, especially through the commercialization of NTFPs (see Plate 7). A well-known example, also relevant in several of the projects' case studies, is the commercialization of the Brazil nut in Northern Bolivia (*Bertholletia excelsa* Bonpl.) (Stoian, 2005; Cronkleton et al., 2012). This region produces nearly half of the world production of Brazil nuts, yielding almost 18,000 tons of shelled Brazil nuts per year. In the Bolivian case studies, which included communities composed of settlers and indigenous people, the Brazil nut represented the main component of livelihood for the majority of families – both in terms of inputs (30% of their total working time) and in terms of benefits (40% of their total income). Similarly, the results obtained in IPHAE (2007) for several communities in the Bolivian study area, especially those situated close to the city of Riberalta, showed that the Brazil nut represented up to 70% of family income, generating revenues of more than US$1,600 per year in high-price years.

Another example of the economic relevance of NTFPs for smallholders were the Brazilian families in the PAE Ecuador case study site, where, in addition to Brazil nuts, rubber production was also an important component of families' livelihoods, especially given the minimum price policy of the state (see also http://www.seringueira.com/br/main.html). For a few families, growing timber in agricultural fields and secondary forests and harvesting NTFPs also contributed significantly to income (Hoch et al., 2012). A showcase model was a producer in one of the Bolivian case studies who, with the strong support of a local NGO, succeeded in establishing an agroforestry system with cupuazu (*Theobroma grandiflorum*), which generated almost 80% of the families' annual income. We also found several examples, such as several communal forests of indigenous groups in Bolivia and Peru and in extractive reserves in Brazil, where communities with large areas of primary forest entered into negotiations with timber companies in the area to harvest their timber stocks and thereby gained, at least in the short term, an important source of income (Medina et al., 2009c). We also found that the importance of forest products in the household economy may grow, at least temporarily, in the setting of development projects (Vos et al., 2009). However, although forests display some income potential – especially for the poorest families – and several possibilities for the lucrative marketing of forest products, overall, the project analysis reveals that the financial attractiveness of forestry production is relatively limited compared with other land-use options due to the following factors: (1) higher required input than other land uses, especially in terms of harvesting and transport; (2) relatively low prices for forest products; and (3) significant logistical challenges in terms of processing, transport and marketing. Consequently, the vast majority of producers whom we visited in the case studies were still far from achieving the high revenues calculated from ideal scenarios by several researchers and technicians (Hoch et al., 2012).

The analysis of the case studies in Bolivia, Peru and Ecuador demonstrates that the possibility of generating attractive revenue from timber depends mainly on the logistical context in relation to the market (Robles, in preparation). When the markets were distant and access difficult, or when the market was controlled by timber companies (as is typically the case in frontier regions), in view of the low prices and elevated transport costs, smallholders did not have the opportunity to generate high revenues from selling standing trees, timber logs or pre-processed planks (see Plate 6). Even in the Macas region in Ecuador, which – in comparison to the other study areas – has relatively short distances to markets and a comparatively large stock of commercial timber in the forests, with calculated revenue expectations of up to US$1,300 per ha per harvest, in practice, the families engaged in forest management in accordance with the law only generated a forest income of approximately US$15 per ha per year (Gatter and Romero, 2005). In Macas, and in many other regions, another factor that also limited the potential for generating significant income from timber was the partitioning of properties as population increases. For example, in the Ecuadorian case study by 'Wachmas' with an indigenous 'Shuar' community, the average size of smallholder properties had decreased to approximately nine hectares, with a distinct tendency to decrease in size even more in the next generation.

Another study analyzing eight of the most promising community forest management experiments in the Brazilian Amazon (Median and Pokorny, 2008; Pokorny et al., 2012) demonstrated the discrepancy between expectations and reality of 'Community Forestry' approaches. The analysis reveals that large investments for capacity building, administration and the acquisition of equipment and machines for operations were needed to enable smallholders to manage their forests within the existing legal framework. Depending on the size and complexity of the productive arrangement, technology used and degree of vertical integration, the supporting development organizations invested between US$20,000 and US$800,000 to establish a pilot project. Although heavy machinery and pre-financing harvesting operations represented the greatest chare of costs for larger initiatives, in smaller initiatives, the costs of technical assistance, including the need to hire forest engineers to comply with legal requirements, represented the most important outlay. In addition to these costs, the participating smallholders invested significant time in meetings and training courses and dedicated large areas of land to the projects, resulting in management restrictions and foregone income earning opportunities without compensation. The magnitude of the investments was confirmed for similar initiatives in Bolivia, Peru and Ecuador, although those countries had slightly lower personnel costs due to lower salary levels compared with Brazil (Medina et al., 2009c; Vos et al., 2009; Robles, in preparation).

One of the most important results of the financial assessment is that, despite the significant external support and the proven ability of smallholders to adopt the proposed technical and managerial production models, the analyzed 'Community Forestry' initiatives showed significantly lower levels of productivity, ranging from 25% to 75%, compared with the levels achieved in mechanized commercial operations working under professional supervision (Table 2.2). Even the larger initiatives most compatible with the proposed Reduced Impact Logging technology packages showed lower levels of productivity than commercial enterprises as a consequence of the social characteristics inherent to Amazonian smallholders (Pokorny et al., 2012). In contrast to enterprises aiming to maximize profits, the smallholders in the case studies were more interested in maximizing employment opportunities. Instead of accumulating capital for investments, they preferred to distribute eventual profits. They also showed resistance to adopting the proposed technologies and specializing in new forest activity and instead continued to value their traditional practices and a wider array of productive activities with more flexible working agendas. These people are accustomed to working on a different time scale, as they had many subsistence and other needs to satisfy, and they could not and did not want to dedicate eight hours per day to a single outlet. Finally, the more egalitarian organization of work and social demands of their family clans affected the performance of operations compared with the vertical hierarchy of commercial enterprises. These phenomena create competitive disadvantages for smallholder operations in markets, which are in turn aggravated by the prevailing public-sector policies (Pokorny et al., in press b). For example, subsidies and incentives offered to the agricultural sector, based on allocation criteria and procedures, were much more readily (sometimes even exclusively) accessible to large landowners and companies that were able to present collateral for loans.

As a consequence of these competitive disadvantages in adopting the proposed technologies for the commercial use of timber, even in the Brazilian case studies – which presented several of the most promising experiments from across the region – the costs of production were high when compared with the timber companies'

Table 2.2   Productivity of timber harvesting operations in relation to scale and organizational approach

	Size and initiative			
	'Community Forestry'			Enterprises
	Mini	Small	Large	Large(1)
	Oficinas Caboclas	*Mamirauá, Pedro Peixoto, Boa Vista dos Ramos*	*Ambé, PAE, Porto Dias, Costa Marques*	*Cikel, Juruá, IBL*
Demarcation (ha/day)	2	3	9	18
Inventory (ha/day)	2	3	11	12
Tree felling (m³/day)	<1	14	40	55
Skidding (m³ per day)	< 1	3	59	75

(1) Data from Pokorny and Steinbrenner (2005)

costs. The identified production costs ranged from US$15 to US$50 per m³ for round wood, US$350 to US$420 per m³ for boards processed with chainsaws and US$190 to US$600 per m³ for boards produced with portable saw mills in the forests (Medina and Pokorny, 2008, 2011). In smaller initiatives, the proportion of technical support and administrative costs were exorbitantly high in relation to the benefits generated. For initiatives processing round wood with chainsaws or portable saw mills, machinery costs were the most significant costs, and taxes on processed wood also represented major cost items. In larger initiatives with mechanized operations, heavy machinery used for road construction, skidding and landing operations constituted the major proportion of total costs. Initiatives attempting to commercialize their products directly to external markets suffered from elevated administration and management costs. Due to the high costs of production, profits were at best modest (Figure 2.3), although all initiatives received relatively attractive prices for their timber from non-local buyers due to the facilitation of the negotiations by the accompanying development organizations, mostly internationally linked NGOs.

Despite the generally low level of local salaries, only the larger initiatives with a low level of vertical integration (Ambé, Costa Marques and Mamirauá) managed to cover labour costs and, in a few cases, generate meagre profits. All other initiatives were unable to even cover the operational costs of logging. However, the initiatives of Oficinas Caboclas, Pedro Peixoto, Porto Dias and PAE at least managed to pay some of their local labour costs. Furthermore, the expectation of enhancing income by adding value from further processing of timber or production of manufactured products was not fulfilled even though attractive prices were paid for the products, mainly because the production costs increased disproportionately as a consequence of more complex administration and the need for essential investments in machinery and equipment. Although FSC certification has permitted access to a wider range of markets, the costs associated with FSC certification were found to be disproportionate in relation to the monetary benefits generated, unless, of course, the costs were met or partially subsidized by external funding sources. Initiatives seeking to add value to timber products also suffered from a higher proportion of fixed costs and a continuous need for capital to pre-finance operations. This situation seriously affected the flexibility of operations while also increasing pressure on production. Families criticized the high degree of risk associated with the need for heavy investments, the need to cover the running costs and, most importantly, the significant time lag before obtaining a return on the investment. In the PAE Ecuador case study in Brazil, for example, families waited nearly two years before receiving any revenue from their first timber harvest (Medina and Pokorny, 2008, 2011).

The evaluation of the potential of forest plantations and agroforestry for smallholders similarly revealed limited financial viability for the families, due to the requirement for significant investment with high associated risks (Hoch, 2009). Although the analyzed case studies were locally thought to be 'promising' and more successful than the majority, even here, only a few smallholders managed to successfully produce and market their plantations' products, and those who

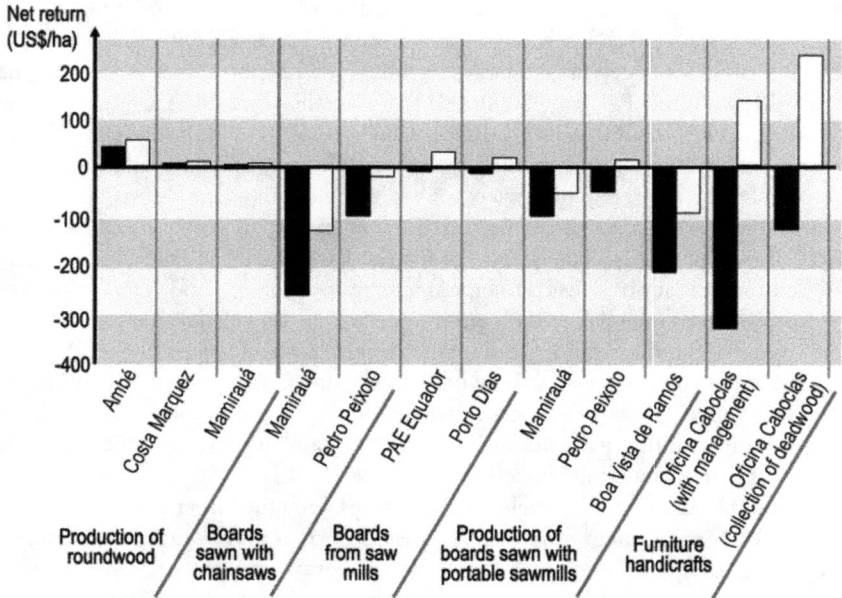

Figure 2.3   Net revenues for the final products of community forest management initiatives in the Brazilian Amazon

did achieved outputs that were generally less than 30% of their initial expectations. Approximately 60–90 working days were required to establish one hectare of tree plantations. This labour, combined with the planting material, easily generated costs of more than US$1,000 per ha. In addition, during the first three years, a further US$50 to US$200 per ha were invested in weeding. Costs for harvesting and transport were significant, although these were more rapidly repaid. Nevertheless, the smallholders, due to a notorious lack of liquid capital, tended to avoid even these costs and, for example, sold standing timber or logs at the farm gate although the prices were significantly lower there. For example, farmers in Peru growing bolaina (*Guazuma crinita* Martius) preferred to sell the standing trees for approximately US$1.7, whereas they would have been able to gain US$3 for the same tree when sold as round wood and even US$6 when processed as boards.

Our field analysis confirmed these findings, as smallholders frequently commented in the interviews that higher prices do not compensate for the higher input costs associated with transporting, processing and harvesting. Another problem arose from the fact that many of the plantations in the case studies were badly affected by drought and fire. The average annual fire risk was calculated to vary between 0.5% and 2.0% per year. Over a 30-year production cycle this represents between 15% and 60% probability of fire affecting a plantation, an assessment comparable to the results of other studies conducted in Panama and

Brazil (Simmons et al., 2002). Plantations also suffered from floods, land invasions, fungi, insects, rodents and occasionally browsing animals. Weeds also significantly hindered the early development of plantations.

The situation was slightly different for agroforestry systems and perennial crops. Although establishment costs were 50–100% higher compared with those of pure tree plantations, and despite the need to reinvest up to half of the generated annual income to cover maintenance and harvesting costs, the experiments analyzed herein showed that net profits could be generated when attractive markets were accessible. For example, we observed one fairly successful initiative implementing cupuazu (*Theobroma grandiflorum*)-based agroforestry systems in northern Bolivia, with annual profits of between US$300 and US$900 per ha. However, during the first six to seven years, this initiative required the investment of approximately 200 working days. Under these conditions, only approximately 15% of the producers initially contacted by the supporting NGO had the capacity and interest to successfully implement and maintain the systems. However, the general existing economic potential of perennials shown by our studies was confirmed by the fact that, in several areas in the Amazon, especially in regions characterized by good infrastructure and fertile soils, certain cash crops such as cocoa (*Theobroma cacao*), coffee (*Coffea arabica*), açaí (*Euterpe oleracea*) and oil palm (*Elaeis guineensis*) have become foci of development (Fujisaka and White, 1998; Shanley et al., 2002; Brondizio, 2004).

In general, the analysis confirmed that there was significant external input for all forestry development initiatives observed in the project. Even in plantation programmes where support was mostly limited to the distribution of seedlings and extensive technical assistance during the initial planting phases, Hoch et al. (2012), in line with similar studies (Current and Scherr, 1995; Sunderlin et al., 2005), calculated costs of US$500–3,500 per ha. Costs for agroforestry projects were significantly higher, as the supporting organizations usually provided more intensive support, which often included the establishment of demonstration sites and training centres, training courses and field visits over relatively long periods. In an attempt to compensate for the absence of local markets, in several of the analyzed cases the supporting organizations themselves became involved in commercialization and invested significant resources in processing facilities, organization and logistics.

In summary, it can be stated that the production and profitability of tree and forest management were generally much lower than initially expected by the development organizations involved (Hoch et al., 2012). In view of the extensive investments necessary, the smallholders (similar to large plantation companies) were interested in making plantations only in areas with better growing conditions – including fertile soils, existing infrastructure and good possibilities for commercialization (Pokorny et al., 2010a). In other words, areas potentially suited for forestry were also suited for other profitable land uses, particularly agricultural production. For the recovery of degraded areas, however, due to sub-optimal production conditions, smallholders generally favoured the regeneration of secondary vegetation growth rather than investment in forest plantations.

Obviously, the commercial potential of trees and forests compared with alternative land uses is restricted. Additionally, the great expectations of profit from the improvement of products and more active participation of smallholders in global market chains were not supported by our findings. In general, the smallholders preferred to produce and sell their products in raw (not processed) form. The principal problem with verticalization occurred because the costs related to the necessary investments for the processing and refining equipment could only be recovered when the production fully utilized the established capacities of the machines. However, achieving full capacity required sophisticated logistics and the establishment of organizational structures and business skills such as those practised by professional enterprises, as proposed by Donovan et al. (2008).

Keeping in mind the potential limitations of the proposed tree and forest management schemes, it has recently been proposed, with increasing intensity, to use the opportunities for payments for environmental services and carbon sequestration that have emerged in discussions related to initiatives for Reduced Emissions from Deforestation in Developing Countries (REDD+; Engel et al., 2008); such payments could ensure the financial attractiveness of local forest management and thus attract the interest of smallholders, particularly for the sustainable management of natural forests. However, in spite of the potential for new sources of income, it is important to keep in mind that the structural difficulties of forest management remain with this approach. Additionally, it is noteworthy that, thus far, the proposed payment models are rarely attractive for smallholders because they entail significant bureaucracy and high initial costs and are consequently more affordable and feasible (as before) for the more capitalized producers and entrepreneurs (see Box 2.2) (Pokorny et al, in press a).

Box 2.2 *Difficulties faced by smallholders in accessing carbon markets*

The projects that are currently being negotiated in the official carbon market, such as voluntary agreements, all follow complex methodologies and regulations to prove that carbon credits are being generated, including requirements such as the following defined by the Unit Nations Framework Convention on Climate Change (UNFCC) (see http://cdm.unfccc.int/methodologies/index.html):

1 Defining the exact boundaries of the project area, excluding all areas that will not be considered;
2 Additionality: payments for reforestation are only possible for areas that would remain degraded in the absence of paid incentives, whereas payments for avoiding deforestation are only possible for areas that would likely be clear-cut in the absence of the paid incentive;
3 Leakage, or demonstrating that the destructive activities avoided through carbon incentives will not simply be displaced to other areas;

4 Calculating carbon sequestration in accordance with externally defined mathematical models;

5 Monitoring, from the start of the project, the establishment of reforestation, forested areas and areas of forest management.

These requirements are not very compatible with the realities of smallholders in the Amazon, so local families will have enormous difficulties in accessing the emerging payment schemes. This problem is further aggravated by the fact that smallholders must compete with largeholders or capitalized actors in providing the most cost-effective option to the donors interested in avoiding deforestation. Thus, in the context of the Amazon, where largeholders contribute much more to deforestation than smallholders, focusing on smallholders is much less attractive when taking into account additionality. Moreover, the small sizes and large dispersion of smallholders' properties considerably increase the transaction costs for such schemes. Realistically, the formal need to apply quite rigid procedures for the verification of carbon credits leaves little possibility of properly considering the realities of smallholders. In the long run, the need for continuous monitoring and standardized reporting tends to diminish the flexibility of these models. Consequently, the new carbon markets, similarly to other externally dominated procedures relying on standardized approaches, will have only a limited utility for smallholders, even considering the efforts to guarantee local benefits by establishing social safeguards in the payment schemes (Lawlor and Huberman, 2009).

Smallholders' capacity to utilise available options

There is a wide consensus that smallholders have a disproportional potential in contributing to sustainable growth, as more is produced per hectare on small farms than on large farms (Cornia, 1985; Heltberg, 1998). In fact, historically, few countries have ever achieved economic growth without a substantial growth in agriculture (Hazel et al., 2010). Despite this, however, there is a general assumption that a profound modernization of smallholders' social and production systems is necessary and that they should be replaced with models and technologies defined by competent and qualified experts (Wiggens et al., 2010). Thus, development organizations and governments commonly assume that smallholders must modify their current practices, cultural organization and working institutions to adhere to these externally defined models so that they can manage their resources more effectively and better benefit from emerging market opportunities (Pokorny and Johnson, 2008a, 2008b; Pokorny et al., 2009, 2010b, 2012, in press b). Against this backdrop, this section intends to show from the findings of ForLive that the smallholders in the Amazon, regardless of whether they are settlers, traditional communities or indigenous groups, are willing and have the capacity to effectively identify and utilize possibilities offered by the market, partnerships and credit programmes to improve their situation. However, in the case studies, it became obvious that families were extremely pragmatic in exploring these possibilities (De Koning, 2011). In accordance with their interests and priorities, they put their interest in social reproduction and risk avoidance at the centre of their decisions. The findings from the case studies suggest that this capacity for pragmatic utilization of emerging options can be interpreted as an inherent characteristic of Amazonian smallholders.

Markets

The continuously advancing frontiers in the Amazon drastically changed the situation of many smallholders (see Figure 2.4). In particular, markets became more accessible and offered better possibilities for commercializing locally produced goods, including timber and NTFPs. The analysis of the case studies clearly proved the capacity of producers to react flexibly to emerging markets and changing market prices. One impressive example of this flexibility was observed in the cultivation of fruit trees. Traditionally, Amazonian families cultivate fruit trees in home gardens, principally for their own consumption (Wiersum, 2004). However, in these case studies, when attractive markets emerged, families began to massively invest in fruit production to sell on a larger scale. When the increased demand was accompanied by an attractive price level, many producers strongly increased their production using low-input strategies for the establishment of fruit tree plantations, as was the case with açaí (*Euterpe oleracea*), camu-camu (*Myrciaria dubia*) and buriti (*Mauritia flexuosa*) (Hoch et al., 2009; Garcia et al., in preparation). For example, in the case of açaí, producers in the eastern Amazon managed to earn a yearly net gain of approximately US$700 per ha by expanding cultivation (Soares, 2008).

Another example demonstrating the pragmatic utilization of opportunities was the frequently observed practice of planting rapidly growing pioneer tree species in agricultural cultivations of perennial crops. In these cases, during the preparation of their fields, the smallholders simply maintained the regularly occurring natural regeneration of these species, notably paricá (*Schizolobium amazonicum*), a flowering legume, laurel (*Cordia sp.*), pau-de-balsa (*Ochroma pyramidale*), bolaina (*Guazuma crinita*), bacurí (*Platonia insignis*), pigüe (*Pollalesta discolor*) and many others. This strategy generated a commercial product with little to zero cost in a short time. In the extreme, a family observed in the ACAPP case study in Ecuador managed to gain approximately US$2,000 per ha by selling timber from pigüe trees grown from natural regeneration, which occurred after temporary agricultural use of the area. The financial viability of this production system, however, was limited to those areas with sufficient natural regeneration and proximity to existing roads, so smallholders would incur only minimal costs for plantation establishment, maintenance and transport (Hoch et al., 2009) (Table 2.3).

It also became obvious that nearly all families in and around the case study sites eventually utilized the existing possibilities of selling trees and timber, mainly

Table 2.3   Costs and benefits for successful producers in two initiatives to promote the commercial production of trees (Hoch et al, 2009)

	Plantations of pioneer tree species with cutting cycles of 4–7 years (pau-de-balsa in Ecuador or bolaina in Peru)	**Agroforestry systems with cupuazu in Bolivia (calculated for the first 10 years)**
Financial return originally expected by the supporting NGO	US$3,000–6,000	US$5,800
Value of the received subsidies (technical assistance, seeds, legal documentation)	US$350	US$1,150
Production	Once every 4–7 years	Yearly, beginning with the fourth year
Total return	US$1,400	US$2,200
Annual return	US$200	US$220

in an informal manner and often on the basis of informal agreements with small logging companies (Medina et al., 2009b, 2009c). When conditions were favourable, as in the sites around the city of Macas in Ecuador, with minimal distance to the market and low competition from professional loggers, the farmers invested in the purchase of chainsaws to process timber into planks and bought horses to transport the planks out of the forest to the roads. Through these investments, the families increased their net income from timber sales by more than 100% compared with the commercialization of standing trees. That said, in the majority of the analyzed contexts, this opportunity did not exist, mainly because an aggressive timber sector was interested in systematically exploring the timber stocks in smallholders' properties. In several countries, such as Bolivia, this situation was worsened because the use of chainsaws to process the timber in the forests was prohibited by law, which strongly affected the price on informal markets for timber originating from small-scale operations.

In line with the International Tropical Timber Organization (ITTO) projections for the development of timber markets (http://www.itto.int), the analysis of the case studies suggests the existence of two important long-term trends: (1) a decrease in timber stocks due to the continued transformation and degradation of primary forests, and, consequently, (2) a continued financial attractiveness of timber harvesting from natural forests due to the scarcity of the most desirable species and because of the opening of markets to new tree species, including the rapidly growing late-pioneer species that are highly abundant in secondary forests (Silva et al., 1995; Schulze et al., 2008). These trends were particularly apparent in the Ecuadorian study area around Macas, where the depletion of timber from traditionally high-value tree species in accessible regions ensured a market for almost all other suitable tree species, including rapidly growing trees such as those of the genus *Cecropia* (Robles, in preparation). Thus, the price for timber from *Cedrelinga catenaeformis* in Ecuador reached US$16 per m³, whereas the prices in those study areas still covered by dense primary forest was much lower (e.g. US$3.9 per m³ in Riberalta, Bolivia, or US$6.7 per m³ in Pucallpa, Peru).

We also found several examples in which increased production of a certain product and the corresponding increase in supply provoked a decline of prices. This situation was observed in Buen Futuro (Bolivia), where families invested in plantations of urucú *(Bixa orellana)*, and in Chinimbimi (Ecuador), where families produced oranges. In both cases, the smallholders were encouraged by development projects to produce these products, but the prices then decreased significantly after the market became flooded. The case of palm plantations observed in the Bolivian study area by Henkemans (2001) can be taken as another example illustrating the risks related to investments targeted on the production of one specific commodity; it also demonstrates the flexibility of producers in reacting to this type of situation. In this last case, when the time arrived to commercialize the palm products from the expanded plantations, Brazil closed the market to Bolivian palm oil to protect their own producers. As a result, the price dropped dramatically. To make the best of this situation, the majority of Bolivian producers then utilised the palm fruits – high in nutritional value – to feed pigs, for their own consumption, and for commercialization in local markets.

We found several other examples in our case studies in which the expected benefits from the smallholders' investments in the production of a specific commodity for markets did not come to fruition. Most frequently, these experiments were promoted by development organizations or public credit programmes and often targeted national or international markets. In view of this uncertainty related to market-oriented investments, it is understandable that smallholders generally favour diversified production with a larger array of products for consumption and markets, thereby avoiding risk, although sometimes at the cost of possible profits (Perz, 2005). Even smallholders who successfully aligned their production systems to use emerging market opportunities tended to maintain their traditional agricultural production to ensure basic food and income to sustain their family. This wariness of smallholders regarding risk and their desire to maintain a stable foundation for their livelihood on one hand, and their capacity for learning on the other hand, also became apparent in those case studies of families participating in development projects. In such cases, the families were often highly ambivalent, as the projects were often unable to meet the initial high expectations (Pokorny and Johnson, 2008a). In the aftermath, the families showed greater resistance to completely following the proposals of experts and technicians. In line with the observation of Rogers (2003), producers generally preferred innovations that afforded the opportunity to experiment and do trial runs before their full adoption.

Another completely different response to the erratic history and changing markets for agricultural and forestry products, which also dramatically demonstrates the flexibility of smallholder families, is the phenomenon of migration. In Bolivia and Ecuador in particular, a large proportion of families migrated to other places that offered (or promoted) better options and opportunities for work. In accordance with Wunder (2001), 'the rural-urban migration path' is likely one of the most successful strategies of the rural poor to become better off. In the region of Macas, for example, the financial contribution to household income of migrants working mainly in the US or Spain was significant, and in many cases it removed the need to produce (Vos et al., 2009). Recent population dynamics (Hecht, 2007, 2010) even suggest that migration in the current Amazonian context has become an intrinsic part of the livelihood strategy of a vast proportion of the rural poor.

Partnerships

The various efforts of governmental and non-governmental organizations to support local development represent worthwhile options from the smallholders' point of view. The analysis of the smallholders in the case studies, who were sometimes strongly supported by initiatives of governmental and non-govern-mental development organizations, revealed two important insights: first, that families were eager to seize these opportunities, which is indicated by the fact that, regardless of the initiatives' content, the families almost always accepted collaborative proposals, typically investing a large of amount of effort and work into these partnerships; and second, that much more than the official objectives of the initiatives – almost always set by the supporting organization – the main motive for collaborating was the expectation of receiving indirect benefits from the new partner. In extreme cases, the producers, for strategic reasons, even accepted proposals that conflicted with their preferred habits to access the benefits of interest (Medina et al., 2009b).

In all study areas, we found families collaborating with external development initiatives with the initial aim of indirectly exploring emerging opportunities for benefits, including benefits that were not necessarily related to the official agendas of the projects. As an example of this pragmatic handling of external partner-ships, one can look to the study of the Extractive Reserve PAE Ecuador in Acre, Brazil. Here, families traditionally harvested NTFPs, mainly rubber and Brazil nuts, and invested in cattle ranching in their individually managed properties. For several years, they have also participated in an initiative for sustainable timber management that was strongly supported by State of Acre and a number of different national and international NGOs. Interviews revealed that the majority of the families did not perceive the promoted approach of timber harvesting on the basis of legally authorized management plans as actually being attractive and meaningful; the majority of these families would have preferred to use their forests in accordance with their traditional management schemes, which focussed on NTFPs. In reality, these families decided to collaborate with the external organi-zations merely for strategic reasons, as they understood that such collaboration could help them gain access to other benefits from the external organizations with 'real' local relevance, including credit, subsidies, investment in infrastructure, health and education. One producer stated that 'they [the supporting government agencies and NGOs] can do with our forests what they want, as long as this guarantees a good relationship of receiving support'. However, this hidden agenda of the local families continuously provoked conflicts with the supporting organizations, as the locals constantly attempted to minimize their input into the external project and systematically used any possibility to adapt the project to their own interests.

Another example of smallholders adjusting the agenda of external support initiatives was observed near Madre de Dios in Peru. Here, the colonist family took advantage of the seedlings distributed within the governmental programme 'Canon de Reforestación' in Peru and established over ten hectares of forest plantations and agroforestry systems. However, they did not do this primarily for the purpose of production, but rather for the possibility of formalizing their land tenure situation with governmental support. The family also diverted the technical advice of the programme regarding the choice of species and planting design to fulfil their own preferences. In a similar vein, a producer in the province of Vaca Diez in Bolivia systematically used accessible programmes and projects to finance the establishment of six hectares of agroforestry systems in line with his own ideas instead of following the technical advice of the financing support organization (Hoch, 2009). However, these cases in which producers successfully managed to use funding and technical support from externally driven initiatives to achieve their own priorities were relatively rare within the analyzed sample, mainly because the external support was typically given on the condition of 'correct' implementation of the guidelines and technologies, which was usually controlled, although often not very effectively. Chapter 6 will show that this inflexibility in development programmes and projects can be interpreted as one of the major reasons for the limited success of many initiatives.

By exploring the Bolivian case study of an indigenous community practicing collective forest management, De Koning (2011) demonstrates the capacity of local families to strategically select appealing elements and ideas of external proposals and recombine them with their locally embedded institutions. With the support of a NGO, the community established a formal management plan and a local association to organize its logging activities. This process subsequently became central to the community's efforts to claim legal ownership of its traditionally managed land. During the course of the initiative, the local forest association gained importance as a social network that, for example, provided support to the families in case of an emergency. Moreover, the scope of the association expanded to other areas, such as the organization of collective agricultural production. A similar situation was reported by Campos (2009) and Nalvarte (2011) in their case studies of an indigenous community near Pucallpa in Peru.

Under specific conditions, mostly characterized by a combination of external threats by capital-endowed actors interested in the smallholders' land or resources and the effective local organization of families based on autonomous interests and endogenous relationships, local grassroots organizations managed to instrumentalise external actors such as NGOs to effectively influence public policy in accordance with their interests (Medina et al., 2009b). A study investigating the possible ways in which smallholders could benefit from their forests in agricultural frontiers (Medina, 2008; Medina et al., 2009a, 2009b, 2009c) showed that, whereas discourses used by NGOs, governments and loggers influenced the ways in which communities managed their forests, several communities were successful in using these discourses to achieve legitimacy and recognition of their demands. In fact, starting in the 1990s, environmental organizations began to describe local communities as a repository of valuable traditional knowledge that is fundamental for the sustainable management and the conservation of Amazonian forests (Schmink and Wood, 1992). Adopting this conservationist discourse, several communities managed to establish alliances with environmental NGOs. In all study areas, we found that these alliances then resulted in significant gains by the communities, securing their rights to their land and natural resources. For example, in Porto de Moz, Brazil, local grassroots organizations representing the traditional communities living in the region had long asked the federal government for recognition of the communities' customary rights but little was achieved. However, after the communities began to align themselves with external support organizations, in particular by establishing an alliance with Greenpeace, the process advanced significantly because this alliance opened the door to increased bargaining power with the government agencies. This leverage resulted in the creation of the 'Verde para Sempre' extractive reserve, with 1.3 million hectares set aside in the year 2004. Similarly, in the case of the neighbouring extractive reserve 'Riozinho de Anfrisio', the local communities achieved the formal demarcation of the lands they had used for generations only after they formed alliances with NGOs (Pacheco et al., 2009; Weigelt, 2011).

Two other examples from Bolivia demonstrate the successful results of this partnership strategy. In northern Bolivia, NGOs played an important role in including a minimum area of 500 ha for smallholders in the regulations regarding the recognition of Community Lands (Terras Comunitarias de Origen; TCO), thus providing access to areas of up to 500,000 ha for almost 4,000 inhabitants (De Jong et al., 2006). In Bolivia, indigenous groups also managed to form well-structured representative organizations at the regional level, such as CIRABO (Central Indígena de la Región Amazónica de Bolivia) and CIDOB (Confederación de Pueblos Indígenas de Bolivia), and they were able to establish strong institutional relationships with NGOs such as CEJIS (Centro de Estudios Jurídicos e Investigación Social) and governmental organizations (Vice-Ministry of Land), thereby successfully influencing the formulation of the new national constitution and catalyzing land regularization to receive formal land titles for their territories (Vos et al., 2009).

Loans and remittances

In the selected study areas, the existing governmental credit programmes generally targeted actors who previously disposed of a certain capital stock and engaged in commodity markets. In all the study areas, due to inherent bureaucracy and the legal and institutional requirements of banks and governments, it was extremely difficult for single families to obtain loans. Those few programmes explicitly designed for smallholders were often not for individuals but for organized smallholder groups such as associations. The high number of local associations found in the region demonstrates that the families generally attempted to fulfil this requirement defined by governments, banks and NGOs to receive support. In fact, a significant fraction of the local associations identified during the project were exclusively created to provide the opportunity to benefit from governmental programmes. Consequently, many of these local associations suffered from a lack of institutional structure and legitimacy and often disbanded after the receipt of the envisaged subsidy.

In the study areas, the credit lines targeting smallholders, if existent at all, were generally linked to the transfer of technologies propagated by experts and research organizations. Loans were often accompanied by a public system for rural extension with an exclusive focus on the transfer of the sponsored technologies (Pokorny and Johnson, 2008b). With this approach, it became apparent that the credit lines were not very applicable to the specific conditions of the smallholders' reality. Sometimes the financed technologies were inadequate or irrelevant. For example, in the case of the Fundo Constitucional do Norte Especial – one of the largest credit lines for colonists along the Transamazon highway in Brazil – the bank required the beneficiaries to establish production systems that were not very suitable for the specific environmental conditions of the region. Although public credit systems became significantly more flexible over the last several years, the majority of smallholders in the study regions still had few chances to overcome the immense bureaucracy and fulfil the formal requirements. In many cases, neither private nor public banks accepted mortgages on small farm properties as collateral for loans, either because the land tenure was still not formalized, because the value of the land was too small to guarantee the loan or because the families lived on commonly or publicly owned land in which the individual smallholding would not be permitted by law to be impounded. This occasionally caused serious difficulties for families searching for capital for investments. For example, in the case of the agricultural cooperative Cooperativa Agricola Integral Campesina (CAIC) of Riberalta in Bolivia, which was involved in the commercialization of Brazil nuts, the families, for formal reasons, were not allowed to apply for bank loans or credit. Thus, they were forced to secure private loans with high interest rates of up to 10% to finance the costs of harvesting, processing and commercialization required for international markets. This resulted in immense costs for the families, requiring nearly one year for the members to receive the returns needed to repay the money invested. Another example is the Asociación de Productores y Productores Agroforestales (APPA) in the same region. Here, local families were able to gain approval for a relatively large project aiming to improve the processing and refinement of their products generated in innovative agroforestry systems. However, they ultimately did not receive the funds because the tenure rights of the members' land, although legally acknowledged, lacked formal documentation.

In the studied countries, the majority of existing credit lines were geared mainly towards smallholders located in those regions with greater agricultural traditions and infrastructure rather than those in the agricultural frontier that characterizes the Amazon. Consequently, the credit lines were poorly adapted to the realities of Amazonian smallholders, who generally had a lower level of mechanization and whose economic activities were diverse, including for example the harvesting of forest products. Moreover, the continental extension of the Amazon has created places characterized by little institutional presence and scattered settlement schemes, which causes frequent difficulties in loan repayment and provokes enormous structural complications for effectively establishing financial programmes for smallholders. Against this backdrop, the high administration costs for small loans make the development of credit lines adequately tailored to smallholders unlikely. That said, perhaps more important than the lack of access, the principal difficulty encountered by the smallholders was the type of technologies promoted by external programmes, which were generally rigid and rarely considered the demands, interests and capacities of the smallholders. Despite these barriers and incompatibilities, all the families observed in the case studies continued searching for any financial opportunity to make investments, given their lack of liquid capital necessary to improve their production systems.

We also found several credit programmes that better served the needs of smallholders. Specifically, this included programmes based on the principle of solidarity credit, run with a capital stock that is permanently refinanced by the smallholders themselves. For example, the Programa Crediamigo administered by the regional development bank Banco do Nordeste in northeastern Brazil and the Institute for Development of Small Production Units (Instituto para el Desarrollo de la Pequeña Unidad Productiva; IDEPRO) in Bolivia both offer credit for small-scale forestry operations.

In several case studies, the families managed to obtain the capital required to improve their production systems through the emigration of family members – often the father or older sons – to urban centres, large cities or even abroad. This was most dramatically apparent in the study area around Macas, where nearly all men left their families to earn money, mainly in the US. When they returned to their families, they invested in agricultural production or forestry – buying chainsaws, forestry equipment, animals (horses, mules, etc) and trucks. We found several examples of returnees who became small forestry entrepreneurs, hiring up to six employees for the production and sale of up to 70 m³ of timber per month. In general, income became an increasingly important livelihood component for families in all the study areas.

Large individual differences in innovation strategies

Although we determined that smallholders generally have the capacity to effectively take advantage of emerging options, as described above, we also found large individual differences among the families. In fact, there was a wide gamut of innovation strategies – ranging from producers who were quite open, had the capacity to negotiate and were eager to adopt new technologies, to those who were much more sceptical and conservative. Many families explicitly preferred to have a quiet life and resisted changing their lifestyle and livelihood for increased production or development opportunities. This phenomenon of varying innovation behaviours has been documented in several studies on the processes of innovation diffusion (Rogers, 2003). As a consequence of the specificities of the innovation, the characteristics of the innovator and development agent, and the contextual conditions, each process of innovation adoption has its own trajectory. Whether the innovation was initiated externally by development organizations or emerged endogenously from the smallholders themselves, there were many different levels of uptake and ranges of success. Nevertheless, the observations in the case studies confirm the existence of large general trends and typical phases. Adapted from Rogers (2003), Figure 2.4 demonstrates the general trajectory observed in the adoption and diffusion of innovation strategies. In the beginning of the diffusion process, there are few people who adopt the innovation. After some time passes, there is increasing adoption of the innovation, and in the final stages, the majority of the members of a social system or community are open to change. Producers who are active in each of these different phases have specific socio-economic characteristics, which correlate in a generalized sense with how they will act at different points within this process. Generally, a social system can be divided into four different categories of innovators: innovators, early adopters, the majority and laggards.

The ForLive study in large part confirmed this typology within the case studies. According to our studies, the innovators can be described as more open and communicative people (including people from outside the community), often with formal education, and generally with enough resources available to take risks by engaging in experiments and investigations. The early adopters had quite similar characteristics and typically belonged to the wealthier bracket of families. However, unlike innovators, they were much more active and socially integrated. Most often they were the leaders, whereas the innovators were somewhat outside

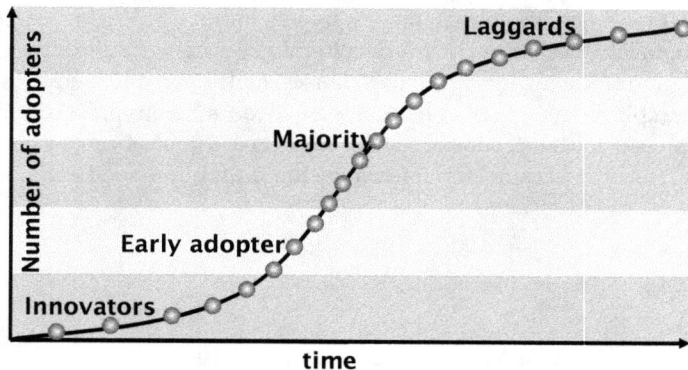

Figure 2.4   Idealized process of innovation diffusion and types of members involved in the social system over time (adapted from Rogers, 2003)

of the social system. The other families, the 'majority', had comparatively fewer material resources (and thus a smaller security blanket) and less communicative capacity. The group of laggards was typically part of the poorest bracket of families and, in the case of traditional settlement structures, e.g. along rivers, it often included families living further away from the centre of the community.

In general, we found that colonists were more open to the introduction of changes in their production systems, whereas, at the other extreme, families in indigenous communities tended to be more resistant to change (Vos et al., 2009). In the case studies of recently established settlements, social cohesion was generally more fractured, and there was little exchange of knowledge between different families. Consequently, there was a large diversity of practices in terms of methods of production; however, this diversity was not effectively used in mutual reflection and learning processes. This observation is in line with an earlier report by Selener (1997), who noted the importance of facilitating exchange among smallholders to overcome possible communicative isolation. In contrast, we found that indigenous smallholders were more homogenous and notably less individualized in terms of their behaviour in response to innovations. In fact, even in study areas where the indigenous families lived in more separated habitats, as in Ecuador and in parts of Bolivia and Peru, their social networks were frequently stronger compared with those of colonists. Generally, there was rather strong integration among families and a strong sense of community, and thus there was an increased potential for the exchange of knowledge. Consequently, we observed stronger barriers to adopting innovations but a rapid diffusion process once the innovations were accepted.

In addition to this temporal dimension of the process of innovation adoption caused by varying degrees of willingness and capacity to adopt innovations, ForLive confirmed differences in the cultural dimensions and attitudes of families regarding innovations, as previously highlighted. Thus, the reflection process with smallholders, experts and researcher abstracted three smallholder cultural types based on empirical observations: first, families who were completely open to a dramatic change in their socio-productive systems and who would utilize whatever possibilities existed to improve their situation (either the richest or the poorest families in the case studies belonged to this type); second, producers interested in improving their production system, although careful to avoid excessively drastic changes; and finally, families not open or prepared to change at all.

3 The potential of smallholders in establishing a stable landscape

Large-scale cattle ranching and family agriculture on small properties are considered the most prevalent land-use models in colonized areas of the Amazon (Fearnside, 2008) although, more recently, well-capitalized actors and agroindustry enterprises producing crops (e.g. soybean, biofuels, and rice) for consumers in distant markets have played an increasingly important role in land-use change and deforestation (Grau and Aide, 2008; Rudel et al., 2009). Much research has been conducted to describe the roles and contributions of the various actor groups to Amazonian deforestation, and there is a general belief that property size has the greatest influence on land-use strategies and, conse-quently, on land-cover change. Thus, the impact of properties of different sizes would be merely a function of scale (D'Antona et al., 2006). ForLive analyzed more specifically the exact ways in which smallholders influence Amazonian landscapes and found that smallholders, through their methods of using and cultivating natural resources, tend to cause less damage to forest ecosystems than large producers such as cattle ranchers or agribusiness, mining and energy companies (Godar et al., 2012b, Pokorny et al., in press b). In agreement with Larson et al. (2006), we also observed that, in the regions dominated by small-holders, the local economy generated more jobs, income from agriculture and forestry was more evenly distributed, and decision-making structures functioned more democratically. Nevertheless, the smallholders also changed the landscape significantly. In particular, on smaller properties, they gradually cut down their forests to make room for crops in the form of small plots and pastures. They also utilized secondary forests of various ages and intensively exploited primary forest remnants. In the case studies, however, the transformation and degradation of forests were usually limited to the area of a smallholder's individual property. Contrary to what was stated by Wunder (2001), this reduced contribution to degradation was not only due to a lack of available labour force and capital, but also appears to be related to the fact that the families were strongly attached to their native lands and had a vested interest in them.

In light of the project's findings, there was a broad consensus that the socio-productive systems of smallholders, in comparison with other land uses

– especially those applied by cattle ranchers and agroindustry – have a greater potential to create a landscape that is environmentally stable while also generating continuous social and economic benefits for the local population (see Plate 9). In this sense, the studies confirmed the observation of Lambin and Meyfroidt (2010) that, thanks to socio-ecological feedback, smallholders adopted more effective and ecologically stabilizing land-use practices because resource scarcity and degradation in ecosystem services impact them directly. In this vein, the findings of ForLive indicate that smallholders' socio-productive systems may become the basis of sound local development when provided with favourable conditions, that is, when they occupy good land, are well linked to public infrastructure and have good social organization.

The contribution of smallholders to deforestation

In all the study areas of ForLive, the smallholders contributed significantly to deforestation. In reality, the conversion of primary forests for other land uses was a part of most smallholders' land-use strategies, except in those cases in which indigenous or traditional communities had huge areas of commonly owned forest land extensively used for NTFP gathering. However, the studies showed that producers cut down forest mainly inside their own lots. The majority of families also showed a preference for maintaining at least a small reserve of primary forest (even when partially degraded by selective logging), especially on relatively larger properties. The majority of families in the case studies showed little interest in adding new areas to their properties; consequently, the forests in collective ownership and the surrounding public forests were minimally impacted by productive activities. This lack of expansive tendencies constituted one of the principal differences between the land-use strategies of smallholders and those of largeholders in the Amazon, particularly capitalized cattle ranchers (Pokorny et al., in press b; Godar et al., 2012b).

This difference was evident in the study by Godar (2009), who analyzed land-use dynamics in four municipalities along the Brazilian Transamazon highway. The study explored the productive activities of colonists and their effects on the landscape over a period of 30 years (1977–2007). During this period, only 16% of the smallholders took advantage of the highly favourable context of 'land grabbing' to increase the size of their property, and the majority of those who expanded their area did so by less than ten ha. Only more recently, some farmers who had successfully engaged in cocoa markets began to systematically acquire land in more remote areas for cattle ranching due to a lack of alternative investment options.

Within individual properties, under certain conditions, the initial process of transforming primary forests for other land uses, although continuous, also occurred significantly more slowly than is often reported in deforestation studies (Soares-Filho, 2006; Fearnside, 2008). In the studied municipalities, for example, during three decades of colonization, the small colonists had deforested on average only 36 ha, corresponding to roughly 40% of their individual properties, which averaged a total of 90 ha. The case studies in Bolivia, Peru and Ecuador demonstrated that this tendency to keep part of the forest intact was even stronger among traditional families and indigenous groups (Robles, in preparation; Vos et al., 2009). For example, one exploratory study by a ForLive project partner, the Committee on Sustainable Development (CDS) in Porto do Moz, Brazil, found that, in 2007, the families in seven traditional communities cut down less than 20% of their individual lots, while there was even less deforestation on communal property.

Against this backdrop, it can be concluded that the contribution of smallholders to deforestation is mainly related to the number of families and the average size of their individual properties. Within their lots, the speed and specific mode of forest transformation depended on various contextual factors such as the fertility of the soils, the pressure from the business sector, the availability of subsidies and technical assistance, markets and their accessibility and the cultural preferences of the families for certain production systems, particularly the degree of affinity for cattle.

However, the expectation that smallholders will not cut the natural forest within their properties is highly improbable in the long run (Billard, 2008). Therefore, the scope and quality of governmental colonization initiatives and consequent governmental support for small-scale colonists likely have the greatest influence on smallholders' real contribution to deforestation, which is strongly related to the rate of continuation and especially influenced by migration. In the study areas, it became evident that a greater rate of migration and a shorter duration of residence on their properties led to a greater impact of agricultural expansion by smallholders. Thus, smallholders' formal or informal colonization of inadequate areas characterized by difficult socio-economic conditions necessarily accelerates deforestation. However, when colonization is allowed in areas with favourable environmental and socio-economic conditions, such as productive land and good access to transport, markets and public services, it has the potential to contribute to the stabilization of landscape dynamics.

However, the analysis of landscape dynamics in the study areas showed that, even in the more extreme cases of unfavourable conditions for colonization, smallholders' influence in transforming the landscape was significantly less than that of other actor groups, particularly cattle ranchers and agroindustry. Table 3.1 shows relevant data for the four municipalities along the Transamazon highway where colonization projects have been established since the beginning of the 1970s. In these areas, although almost 90% of all the individual properties belonged to smallholders and only 3% were occupied by large producers (mainly cattle ranchers), the two actor groups occupied similar proportions of land area due to the large difference in the size of the individual properties. Starting with the beginning of the colonization 40 years earlier, the data demonstrate that each cattle rancher deforested on average one-third of his property, corresponding to 555 ha. Thus, each cattle rancher deforested 15 times as much as one smallholder family. In the study area, the medium and large producers – comprising 12% of all colonists – deforested more than all the small colonists put together.

Table 3.1 Contribution to deforestation by actor type in four municipalities along the Transamazon highway in Brazil (adapted from Godar et al., 2012b)

	Proportion of total number of producers (%)	Average property size per producer (ha)	Average area deforested in individual property (ha)	Total contribution to total deforestation at the municipal level (%)
Smallholders (< 200 ha)	88	90	36	47
Medium producers (200–600 ha)	9	402	174	24
Large producers (> 600 ha)	3	1850	555	29

Similar proportions were observed in the study area around Pucallpa, Peru, where the cattle ranchers and, more recently, palm oil producers deforested approximately 59% of their properties, whereas in the areas still dominated by indigenous groups and smallholders only 10% of the forests had been lost despite a much longer settlement history. The same phenomenon was observed in Bolivia, where according to the FAO (2010), approximately 290,000 hectares of forest were cut each year between 2000 and 2010. In the study area around the city of Riberalta, data from the Forest Superintendence of Bolivia for the years 2004 and 2005 show that only 2% of landowners were responsible for more than 40% of deforestation, whereas smallholders contributed less than 20%. Here, the primary responsibility lay with cattle ranchers and soybean producers, who, driven by the dynamics of international markets and improved roads, expanded their production areas significantly in recent years (Blum, 2009). Several other studies have confirmed this strong influence of large producers on deforestation in the Amazon and in other regions (Fearnside, 1993; Mertens et al., 2002; Pacheco, 2009a, 2009b).

The landscape created by smallholders

As demonstrated above, smallholders tend to convert the forest inside their own property. ForLive found, however, that this activity – under certain favourable conditions – could be part of a transformation process from the originally forested landscape to a cultivated landscape characterized by a dynamic mosaic of different land uses (see Plate 10) that have the potential to ensure environmental stability while effectively providing the goods and services needed by families for their well-being. In this array of land uses, within the integrated vision of the smallholders in accordance with their resource level, the forest was an intrinsic part of agricultural activities and vice versa (see Chapter 2). There were mutual synergies in both the temporal and spatial relationships, resulting in a complex mosaic of different land uses including forests used and managed for different purposes and at different intensities. The diverse elements of these landscapes were created by interventions by the smallholder at different spatial scales (from a few square meters up to several hectares) and different time scales (from a few days up to several years) before and between possible treatments. The project found indications that this temporal and spatial variation has the potential to ensure an ecological basis for permanent use of the land.

Plate 10 was constructed by interpreting satellite imagery for the city of Medicilândia, located on the margins of the Transamazon Highway in the state of Pará, Brazil (Godar, 2009). The image provides an example of the type of landscape resulting from smallholders' conversion of forest to cultivated land. In the Medicilândia example, the military government distributed individual properties of 100 ha to families mainly originating from the poorest regions of northeastern and southern Brazil in the 1970s. The lots were distributed, in accordance with the classical fishbone pattern, along secondary roads perpendicular to the main road. The black lines indicate the boundaries of the individual properties, and the roads appear in white. In addition to the original matrix of primary forest (dark green), which was significantly reduced, the other colours indicate the various types of land use at different times and at different intensities. The pasture areas in red appear quite scattered in discontinuous pieces of reduced size. Thus, grazing areas did not grow into larger areas, which would have had serious ecological ramifications for the land owners, such as a high risk of soil erosion. Cocoa plantations, established on the lots in areas with higher soil fertility and appropriate physiography, appear in blue. The secondary forests were abundant and presented different successional stages (the younger in yellow, the older in light green). The existence of these secondary forests indicates that agricultural use – mainly cattle ranching, performed in the initial settling phases as needed to prove appropriation and use and so receive governmental support – was previously practiced in many areas but abandoned long ago. In other parts of the forest, characterized by shifting cultivation agriculture, forest fallows were used to maintain soil fertility using natural succession dynamics.

The potential of landscapes created by smallholders

As shown above, under favourable conditions, smallholders tend to transform the originally stocked primary forests inside their lots into a spatially and temporally dynamic mosaic of diverse land-use components. In this mosaic, the forest, especially the secondary forest but also fragments of primary forest remnants, play a fundamental role. Together with natural forests eventually maintained in the public areas around the smallholders' lots, the secondary forests generate environmental services that are important at the local and global levels. The project also found indications that, beyond their environmental potential, landscapes dominated by smallholders also had the potential to generate significantly better socio-economic value than the landscapes influenced by larger producers such as cattle ranchers and agroindustry enterprises.

Environmental potential

Many authors have highlighted the positive role of traditional communities and indigenous groups in the preservation of the forests in the Amazon (Schmink and Wood, 1992; Campos and Nepstad, 2006; Chhatre and Agrawal, 2009; Nelson and Chomitz, 2009). Many of these studies refer to the role that smallholders play in protecting environmental conservation areas established in relatively remote regions. The findings of ForLive, however, demonstrate that these observations may also be valid in more dynamic contexts (including settlements) dominated by smallholders. To assess the environmental potential of landscapes influenced by different groups of actors, Godar et al. (2012b) compared landscape metrics (forest fragmentation, forest core area and forest connectivity; see Chapter 1) in four municipalities along the Transamazon highway. One of these munici-palities was dominated by smallholders (Medicilândia), whereas the other three (Brasil Novo, Anapú and Pacajá) were strongly influenced by medium and large producers, mainly cattle ranchers. All the calculated environmental indicators listed in Table 3.2 reflect a strong degree of transformation of the initial forest landscape. However, in the municipality of Medicilândia, where the smallholders had a predominant influence, the indicators were significantly better than in the other three municipalities dominated by medium and large producers. Despite being colonized at the same time as the other municipalities, Medicilândia had more areas of continuous forests, greater connectivity of the forest fragments and, consequently, a lower fragmentation index. We found temporal and spatial continuity between the various components of use, both in the structure and connectivity of the vegetation as well as in the products and services generated.

Although the specific implications of the calculated metrics for biological conservation are not fully understood, the metrics effectively express different patterns of deforestation that have proven implications for the provision of environmental goods and services (Manel et al., 2003; Laurance et al., 2002; Mesquita et al., 1999). Thus, the metrics calculated for Medicilândia suggest that the diverse landscape mosaic created by the smallholders' way of using resources has great potential for the continuous generation of environmental services such as hydrological protection, carbon sequestration, maintenance of soil fertility, seed dissemination, genetic flow of flora and fauna and transit of wild animals. Such landscapes created by smallholders, although they have undergone significant transformation from the original forest landscape, may have the potential to contribute to the maintenance of these environmental functions, especially when compared with the more homogenous landscapes created by the cattle ranchers and agroindustry enterprises.

Support for this hypothesis was also found in the region of Pucallpa, Peru, where, in the zones dominated by traditional smallholders, the forest area was maintained with little fragmentation and high environmental value (see Plate 11). In this region, the traditional manner of using natural resources sustained a complex hydrological system and diverse types of vegetation. In the zones dominated by indigenous groups and traditional communities (right), producers

Table 3.2   Qualitative indicators of the landscapes in municipalities dominated by smallholders versus municipalities dominated by medium and large producers (adapted from Godar et al., 2012b)

	Dominated by smallholders	Dominated by medium and large producers
Municipalities	Medicilândia	Brasil Novo, Anapú and Pacajá
Production type	Agricultural (cocoa, banana and coffee)	Mostly ranching
Area occupied by smallholders (%)(1)	71%	40–50%
Deforestation per person (ha)	6.8	9.6–12.5 ha
Forest core area (300 m)(2)	81	51–78
Connectance (500 m)(3)	48	26–31
Fragmentation of the forest(4)	253	322–474

(1) Producers with an individual property up to 100 ha.

(2) Areas of forest located more than 300 m from the nearest unit of humanized landscape (roads, pastures, etc) with a baseline in 1987 of 100 units; indicates the state of effective conservation of mature forest and its potential for maintained provision of goods and services.

(3) Defined as the number of connections between forest patches with a functional distance of less than 500 m with a baseline in 1987 of 100 units; indicates the degree of spatial continuity in the forest, enabling or hindering different processes (e.g. hydrological protection, seed dissemination and the movement of animals).

(4) Ratio of the number of small and unconnected forest patches between 1987 (= 100) and 2007; for example, in Medicilândia, there were 2.53 times more small and disconnected forest patches in 2007 than in 1987.

have generated only three areas of stronger forest penetration (white circles) over several decades, taking advantage of high soil fertility to perform shifting cultivation agriculture with a low environmental impact. In contrast, in the cultivated areas with a predominance of colonists engaged in cattle ranching, mechanized agriculture and palm oil production, only smaller areas of forest remain, causing huge negative environmental effects. In this area (blue circle), the farmers continued to systematically extend their properties to establish pastures and fields at the cost of the original forest.

Additional evidence for the potential environmental compatibility of small-holders' socio-productive systems comes from the analysis of local practices for timber utilization in the Bolivian, Ecuadorian and Peruvian Amazon (Robles, in preparation). This study shows that, in specific contexts characterized by the absence of excessive pressure from the timber sector, smallholders used their timber stocks in a way that ensured the maintenance of the financial and environmental functions of their forests over time. Typically, the low-input harvesting systems used by families were characterized by the following: a selective use of trees, starting with the tree species of greatest commercial value; a broad spatio-temporal distribution of operations, simulating, to a certain extent, the natural regeneration dynamic of the forest; respect for a minimum diameter of 50 cm dbh (diameter at breast height) for harvesting; processing timber logs in the forest with chainsaws; skidding the boards with horses; and the protection of mature trees that provide seed. The productive sustainability of such local management systems manifested itself in the composition of the tree species in the managed forests. In several of these forests, we found an even larger quantity of commercial tree species *(Aniba sp., Ocotea sp., Dacryodes peruviana, Otoba parvifolia* and *Vochysia sp.)* in diameter classes of up to 60 cm dbh compared with unmanaged natural forests (Figure 3.1). However, it should be noted that this valuation does not consider the danger of the possible extinction of several valuable rare species such as mahogany (*Swietenia macrophylla*) and cedar (*Cedrela sp.*) arising from a focus on harvesting.

This increasing abundance of rapidly growing and commercially valuable trees, which was confirmed for many secondary forests in the smallholders' properties, could make for an economically interesting resource. Commercial pioneer species such as capirona or guayabochi (*Calycophyllum spruceanum*) and bolaina (*Guazuma crinita*) in Peru, and pigüe (*Pollalesta discolor*) in Ecuador, showed densities of up to 150 individuals per hectare. These inventories demonstrated that, in older secondary forests of approximately 30 years of age, up to 80% of the trees could have commercial value, although not all boles were of high quality. In secondary forests, various non-timber forest product tree species were also identified. In the analyzed case studies, for example, the number of Brazil nut trees (*Bertholletia excelsa*) was three times higher in secondary forests than in primary forests (Robles, in preparation).

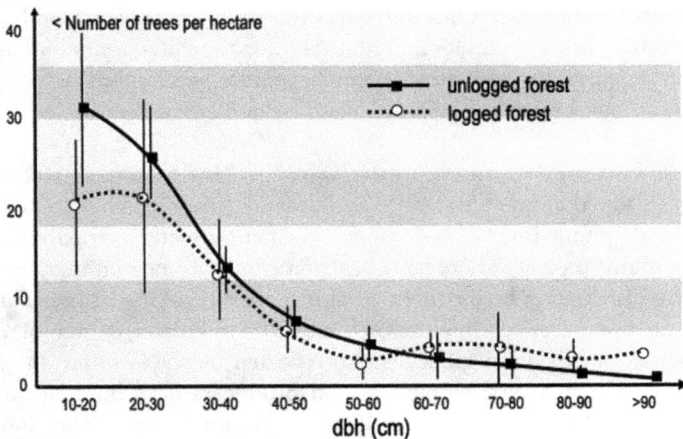

Figure 3.1   Comparison of the dbh (diameter at breast height) distributions of the four most commonly harvested tree species in the region of Macas (Ecuador) between forests harvested by smallholders (n=6) and non-harvested natural forests (n=3)

Further evidence of the environmental potential of smallholders' production systems emerged from a study on the effects of local agricultural practices on soil fertility in the region of Macas, Ecuador (Andersen, 2007). Nutrient analyses of the soils in different land-use systems within smallholder properties indicated that secondary forests, silvopastoral systems and home gardens had soil nutrient levels that were comparable to those of primary forests. Another study (Sato et al., in preparation) of informal logging in the region of Porto de Moz, Brazil showed that, in the informal use and commercialization of timber, smallholders who collaborated with small-scale loggers knowingly took care to leave mature trees that provide seed in the forest to ensure the regeneration of timber species with commercial potential.

Social potential

The smallholders' socio-productive systems, compared with those of large-holders, showed potential advantages not only with regard to the generation of environmental services, but also with regard to the generation of relevant socio-economic benefits at the local scale. This perspective has often been disregarded in studies of frontier dynamics that focused merely on deforestation. However, this perspective seldom acknowledges the role of colonization and the use of Amazonian resources in providing socio-economic opportunities for poor people (e.g. Chomitz and Thomas, 2003; Fearnside, 2005; Margulis, 2003; Michalski et al., 2010).

Against this backdrop, the above-cited study by Godar (2009) on frontier dynamics along the Transamazon Highway also demonstrates that smallholders' production systems, under adequate conditions, have the potential to generate a more regular income for a greater number of families than large-scale production systems. Accordingly, the insights gained in the study areas clearly indicate that large-scale production systems for agriculture, cattle ranching or timber harvesting in the form practised by capitalized actors have little potential to reduce poverty and may even compromise future options for the effective use of the land and resources.

As shown in Table 3.3, in the municipality of Medicilândia – where small-holders were the dominant influence – residents' income per capita and the Human Development Index (HDI) were significantly better compared with the other three municipalities, where large-scale farming was the dominant economic activity. The distribution of income was also more equitable in Medicilândia, as shown by the comparatively higher Gini index. The greater degree of equality between actors in this municipality is also reflected by the fact that the average productivities of employers, employees and producers here differed much less than in the other municipalities. In Pacajá, a municipality highly influenced by large-scale cattle ranchers, employers earned 20 times more than their employees, whereas in Medicilândia, the employers earned only two times more. Additionally, the percentage of the population below the poverty line was lower in Medicilândia than in the municipalities with more large producers. The health and education indices in Medicilândia were also better, therefore demonstrating the potential of smallholders' production systems to contribute to local development. The mortality rate for the population in this municipality was the lowest in the region, although there was only one established public health facility. Similarly, the literacy rate was better, and the dropout rate in primary school was among the lowest in the entire region along the Transamazon Highway.

Trajectory of land-use optimization

Generally, studies on the effects of colonists on landscapes suffer from short observation periods and tend to ignore the history of land use and long-term trends (Ferraz et al., 2009). In fact, frontier expansion is a very dynamic process in which actors deforest at different intensities according to livelihood cycles, market dynamics and production conditions. However, the majority of deforestation assessments rely on static data and do not consider the improved land-use efficiency dynamics of certain actors (Pacheco, 2009a, 2009b). Ideally, both the initial colonization phase and the advanced frontier phases should be

Table 3.3   Municipal indicators of socioeconomic performance

	Medicilândia	Brasil Novo	Anapú	Pacajá	Region
GNP per capita (2004) (USD) (1)	4,449	1,979	3,725	478	1,833
Human Development Index (HDI) (2000)	0.71	0.67	0.65	0.66	0.67
Gini Index (2000)(2)	0.41	0.46	0.48	0.53	0.44
Average productivity (2000) (USD/ month)(3) Employers	890	2,289	1,428	6,976	2,280
Employed workers	445	326	320	364	356
Self-employed workers	651	460	523	387	475
Population below the poverty level (%)(4)	47	44	57	69	51
Total mortality per 1,000 inhabitants (2004)	3.13	3.28	8.62	4.25	3.80
Health facilities per 10,000 inhabitants (2003)	4.4	3.6	6.7	2.5	4.0
Literacy rate (2000) (%)	79	79	73	74	78
Dropout rate in primary school (2004) (%)	17.30	11.00	31.70	21.50	17.83

(1)Exchange rate 1 USD = 2,66 R$ (31.12.2004)

(2)Gini index: measures the degree of income distribution, with a value ranging from 0 (perfect equality) to 1 (maximum inequality) (IBGE)

(3)Exchange rate 1 USD = 1,81 R$ (30.06.2000)

(4)Proportion of persons living in families with a monthly income below half the minimum salary per capita

taken into account for reliable interpretation. Although initial deforestation is often intense, this dynamic can sometimes be reversed during the later stages of colonization (Pfaff, 1999). In this respect, the results obtained in ForLive indicate that the landscape transformation process undertaken by smallholders – when in favourable contexts – may occur in three principal phases:

Phase 1 Deforestation for agriculture, often accelerated by a formal or informal need to prove possession and ownership of the land in order to gain legal acknowledgment, access eventual public support and demobilize potential invaders;

Phase 2 The cultivation of diverse products, generally associated with an expansion of agricultural areas – often including cattle – and frequently fuelled and directed by the type of technical and financial support available; and finally,

Phase 3 The optimization of a production system on the property, contributing to a process of ecological stabilization of the landscape.

Plate 12 illustrates this process of transformation and optimization using the example of Medicilândia, where smallholders settled in the 1970s and have remained ever since. In 1991 (left), a large portion of the colonized area was previously cleared for agricultural activity (red), principally for cattle ranching, subsidized by the government. There were still large areas of primary forest (dark green). In 2007 (right), a considerable portion of the agricultural areas from 1991 were transformed into a mix of agricultural lands, grasslands, younger forest fallows (yellow), older secondary forests (orange) and cocoa plantations (white), often combined with individually grown trees.

Another interesting detail is the visible effect of governmental efforts to protect the indigenous reserve of Arará (southwest of the blue line) during the colonization process. Although this indigenous reserve was previously established in the 1980s, only since 1989, after a number of violent conflicts between the colonists and the indigenous group, did the government effectively begin to control the area and drive out families who were illegally settled in the reserve. Consequently, the deforested areas from 1991 (in red), transformed by invading colonists, had reverted to forest (green) by 2007. Governmental control apparently also reduced illegal logging activities. Thus, by 1991, huge parts of the reserve were affected by illegal timber harvesting, as indicated by the mosaic between light and dark green within the reserve area. In the year 2007, the changed colours in the picture indicate that logging, although still persistent, had decreased considerably, allowing for the recuperation of a large part of the affected forest. Clearly, with enforcement efforts by the government, the reserve proved to be an effective barrier to the process of deforestation, demonstrating the huge importance of effective law enforcement in public areas. However, one must consider that, as a result of this control, many of the colonists who were expelled from the reserve began to occupy other public and private areas without legal reserve status.

In nearly all the case studies, we observed evidence of this process of environmental consolidation through an optimization of the production system, often induced by the limited availability of natural resources on the property. In reality, the shrinkage of forests and the increase in degraded areas on the properties in the long term implies a growing necessity to make adaptations in order to optimize and more effectively utilize the possibly depleted original resources. This process was supported by the smallholders' practice of continuously making small adjustments and engaging in small-scale experiments related to productive options, a capacity previously observed by Biggs (1990). Thus, these latent efforts in the optimization of production systems, observed in literally all the case studies, may be considered typical for smallholders. In practice, however, the full realization of the final phase of optimization was rarely observed. A reflection on the events within ForLive revealed two notable reasons for this shortfall: first, the long duration of this process (up to several decades) against the relatively short time frame of colonization and market contact in the studied cases; and second, the generally unfavourable contexts and the lack of the necessary conditions to facilitate smallholders' exploration of their potential, given the current social, environmental and political conditions that tend to marginalize these people (see Chapter 4).

Based on the insights gained during the project from the case studies and the many discussions with experts and smallholders, we found that a favourable evolution of smallholders' production systems is supported by four main factors: (1) a limited availability of human power and capital, which restricts the possibilities for smallholders to expand their properties; (2) the existence of a set of locally adapted production practices that are compatible with the environmental and socio-economic conditions; (3) the tendency of smallholders to focus on low-risk, low-input activities within their own areas under family control while avoiding sudden, large-scale experiments and dependence on external actors; and (4) a strong cultural link to their land and productive activities. As these conditions were, at least partly, fulfilled in the above-cited case of the municipality of Medicilândia, it was possible to observe this optimization process, at least gradually. Here, the families had several decades to develop and optimize their production systems under more favourable conditions, including comparatively fertile soils, access to attractive markets, security in their land tenure, an elevated level of social organization (often externally catalyzed) and low pressure from the private sector (see Chapter 8).

4 Marginalization of smallholders

The previous chapter demonstrated the potential of smallholders to create land-use strategies with high environmental, economic and social value. Although smallholders do transform primary forest into other landscapes within their individual properties, the smallholders' land-use systems also demonstrate the potential for generating a dynamic mosaic of diverse land-use components, which, as a whole, has a high environmental value and also meets the social needs of the local population. However, we found that this potential is rarely realized in the areas that we studied. In this chapter, we explore the reasons for this unfulfilled potential. Our analysis shows that smallholders are systematically deprived of the opportunity to develop their capacities. The societal relationships that marginalize poor rural smallholders began with the historical colonization of the region and still exert a strong influence on the prevailing development model as well as on smallholders' access to development opportunities.

A brief historical trajectory of 'development' in the Amazon

Five centuries of colonization, settlement and exploration in the region have shaped a societal structure that continues to impact events today. Since the beginning of the colonial period in the Amazon, Europeans have established mechanisms to exploit resources of interest (drugs, gold, sugarcane, etc). In much of South America, society was stratified vertically so that a group of a few elites had control over the majority of the land and resources. Furthermore, after the first contact between the 'conquerors' and the indigenous peoples, the church came to the region to systematically spread the 'word of God' and Christian doctrine, thereby creating a smorgasbord of cultures and production modes that constitute a large part of the societal groundwork.

The invention of private property

Of all the specific historical antecedents of the societal problems found in the region today, it makes sense to specifically highlight the nineteenth century, when many countries started to introduce private land ownership. Simply put, two contexts can be distinguished: in many Andean countries, the liberal reforms of Simón Bolivar to establish a class of independent smallholders promoted the creation of individual properties. However, this reform was enacted by systematically dividing communal and mostly indigenous land, whereas the previously existing feudal domains were allowed to remain (Keen and Haynes, 2009). In other South American countries, the ruling class introduced private land ownership to ensure continued access to resources and the exploitation of cheap labour in response to the ban on slavery. In Brazil, for example, this process was crystallized with the passage of the *Land Law* in 1850 (Wienold, 2006).

Despite their strongly differing intentions, both contexts led to similar consequences, as they both initiated an extensive process of transformation from open public land to privately owned areas in the hands of local elites, including landlords, public officials and merchants. This process was accelerated in the second half of the last century by the prevailing agrarian reform policies, which systematically encouraged the privatization of land in many countries. Although poor rural families had the legal right to obtain property in theory, in practice, the social elites managed to effectively establish mechanisms to limit poor peoples' access to these possibilities with their political influence, corruption of private property legal systems or even physical force against revolting families and their representatives. The elites also often used various appeasement strategies to keep poor land managers in a dependent relationship with those with power. In Brazil, for example, the economic elites invented several different legal ways of 'allowing' the presence of small producers on public or private lands, but the poor families were not granted legal property titles or control of the land and resources that they were managing (Wienold, 2006). In practice, the system was bogged down in the creation of complex legal regulations and procedures that were lengthy, bureaucratic and not easily accessible (Bunker, 1985). In other countries, such as Peru, the government was also greatly influenced by a few powerful 'hacendado' families who created a legally institutionalized tribute system as a way to ensure control over land and people (Keen and Haynes, 2009). Thus, many governments created a type of 'legal allowance' wherein smallholders were encouraged to stay on their lot, but they did not receive an official land title and were thereby subordinated to the interests of the elite. This situation helped to create a social system between rural families and largeholders based on paternal dependence.

Initiatives for the colonization and the exploitation of resources

Since the sixteenth century, the Amazon economy has been structured in accordance with global markets. Local development was either neglected by growth policies or these very policies were designed to favour local elites. Another important process that reinforced the historically paternalistic structure of Amazonian societies was the strategy of many countries to integrate the region into their national economies. After World War II, many Amazonian countries experienced economic growth fuelled by the increasing economic influence of the hegemonic power of the US. This trade dynamic, however, was principally driven by the export of minerals, raw materials and agricultural crops, and it focused on other, more accessible regions of Latin America. The orientation of the region towards the adoption of high-input and large-scale agricultural production systems can be interpreted as the latest version of these policies or extractive cycles. It is noteworthy that the majority of the raw materials and primary goods produced in the region were processed outside of the region, with only a small proportion of the benefits remaining in the region, mostly concentrated in the rapidly growing urban centres (see Plate 8), where the accumulated wealth skews the regional statistics on health, nutrition, education and consumption (UNDP, 2010). This imbalance is one of the reasons for the very low apparent contribution of the Amazon sub-regions to national GDPs. With the possible exception of Ecuador, all the Amazonian countries have developed a strong national disparity between more 'developed' regions (the south in Brazil and the coastal regions in the Andean countries) and the Amazonian regions (Pokorny et al., in press b). Moreover, in the Andean countries, the Amazonian region is also physically separated from other areas by the high mountains of the Andes. In Brazil, the separation was mainly due to the huge size of the Amazon, which, until the 1970s, completely lacked any infrastructural connection to the industrial centres in the south.

In the course of economic growth, efforts aimed at decolonization and change in the power balance provoked many social conflicts. In many of the Amazonian countries, for at least several years, military governments came to political power, resulting in attempts to define development in accordance with the values and interests of their military dictatorships. Several of these military dictatorships, such as in Peru and Bolivia, had, at least temporarily, an explicit focus on a social agenda. However, reforms were often undertaken in a half-hearted manner, as the people in power were often dependent on or strongly connected with powerful oligarchies of elites. Other governments attempted to function more openly in an attempt to maintain their own power and ensure economic well-being in the previously developed regions (Wehrlich, 1978; Ryder and Lawrence, 2000; Giugale et al., 2003; Klein, 2011).

In this context, the Amazon region gained importance mainly for two reasons: first, there was an increasing awareness and demand for the immense store of resources that could be sold to export markets or used to supply the growing industrial centres in other parts of the Amazonian countries; and second, in a related issue, governments became concerned with ensuring possession of the land and resources for geopolitical reasons (Kohlhepp, 1991). The latter issue provoked several territorial, sometimes even violent, conflicts. The intensity and mode of integration of the Amazon into other parts of the country and the exploitation of economically relevant resources differed significantly from country to country. However, in all cases, this process was grounded in the perception that the use of resources in the Amazon consisted of exploiting and cultivating otherwise 'socially useless' forest lands for national benefit.

Ambitious infrastructure projects were begun and, in line with the neoliberal thinking that strongly influenced policies in this post-World War II period, the measures taken often required huge investments. Privatization and private incentives became the means for getting the needed capital and the projects often involved international capital. Consequently, stocks of minerals, oil, forests and gases were often sold or promised in long-term concessions with the simple condition that the private investors pay, or at least participate in, the costs of exploring and exploiting the resource. Huge tracts of land were also distributed, in some cases to small settlers but – again, in view of the 'costs' of establishment – also to capitalized people and companies, motivated by subsidies and transfer payments.

In several countries, this process of Amazonian colonization only began on a large scale in the late 1990s and recently gained more impetus due to large and ambitious national and international infrastructure programmes such as the Initiative for the Integration of Regional Infrastructure in South America (IIRSA), which were then being increasingly oriented towards the Amazon. In other countries such as Brazil, the strategic colonization of the Amazon was initiated in the 1960s and 1970s with the National Plan for the Integration of the Amazon (Kohlhepp, 1991). In this case, one important reason for 'cultivating' the Amazon region was to relieve increasing social conflicts in other regions, where the majority of the land was controlled by a few powerful families. It was expected that the provision of the comparatively bountiful land resources to revolting poor tenants and rural workers could avoid, or at least postpone, agrarian reforms. However, this history is marked by stops and starts due to the political instability of the region. In the 1970s, four years after the National Plan for the Integration of the Amazon was begun, the Brazilian government abruptly changed its strategy – which was initially oriented towards settlements for smallholders – and started giving large amounts of support to enterprises and capitalized actors by investing in the construction of basic infrastructure and establishing special credit programmes administrated through the Superintendence for the Development of the Amazon (SUDAM) and development banks (Treccani, 2008). Consequently, it was the capitalized actors, mostly companies and largeholders from the south, who received a huge portion of the public payments, whereas the support for small settlers migrating from conflict areas and the poor living conditions of their home regions was modest, if it existed at all. Even in the government settlements, families seldom received anything more than the land and, when lucky, some initial support to build houses and buy basic equipment. The imbalance of support in favour of capitalized actors reinforced the social inequality predominant in the Amazonian region throughout its history.

Isolated societies in the rural Amazon

We found that the exploitation of Amazonian resources in the study areas was historically directed by the interests of powerful elites and that this is still currently the case; these exploitation efforts generally neglected the attainment of sustainable economic development for the region, as reflected in the precarious situation of many poor families (Blum, 2009). Typically, these poor families were not active directors of their future, but a group who were, at best, simply disregarded but more often used by the powerful local economic and political elites.

This context was characterized by a lack of state government presence and a poor availability of public services. The resulting regional power vacuum left a high level of informality in which local elites and capitalized international actors had the opportunity to exert their power as they deemed suitable, especially as the frontier reached the smallholders and they became more closely linked with these larger-scale actors.

A well-known example of this type of paternalistic relationship is found in the rubber extraction industry in Bolivia and Brazil, where the 'rubber barons' managed to organize the extraction of rubber from immense areas by exploiting the work of poor families, who they settled there (Dean, 1989). Instead of paying for the raw material and labour in cash, they often established a barter system, in which the land owners provided the basic daily necessities for the families, thereby establishing a system of debt and creating complete dependency. This system, known in Brazil as 'aviamento', was still practiced in more remote areas associated with the rubber economy or in the extraction of several non-timber forest products (NTFPs) and agricultural crops. Many people who migrated to the Amazon have been subjected to this type of paternalistic system. The result is a history of dependence, with smallholders economically dependent on employers who have the power to define the social and political realities of the region, including the use of the land and forest (Reis, 1997).

Historically, local societies were isolated and had little conception of the value of their resources or how to protect their interests. In fact, the majority of families had little contact and communication with outside actors, and they acted fairly independently of the formal legal and institutional settings and national-level politics that were defined in distant urban centres. Instead, these families managed their resources based on their own knowledge and in line with their cultures and traditions. Their management techniques were guided by a trial-and-error approach for practical implementation. As described in Chapter 2, the production systems were generally simple but diversified, showed a low level of mechanization and almost exclusively used the internal labour force of the family. Moreover, due to the smallholders' lack of liquid capital, they avoided larger investments. In more remote areas especially, the natural resource management systems were principally focused on food security and social reproduction and subsistence rather than on income generation, let alone profit. Families in these areas still depended on middlemen or patrons to sell their marketable products, which implied a high dependency on these people, a lack of power to set prices and value resources and ultimately the receipt of low prices. Often, the forests and natural resources outside the properties were extensively used for game and the harvesting of NTFPs, and thereby contributed to the families' subsistence. Due to the low intensity of use and low population density, the landscapes in these areas still held abundant forests of high environmental quality. However, the smallholders living in these areas were generally poor, and their food supply and health situation were often precarious, especially during the rainy seasons, when many of the villages were literally cut off from the outside world.

Paternalistic visions of development

The initiatives for rural development themselves were also part of the historical trajectory of development in the Amazon. Thus, 'development' itself was framed paternalistically (Chambers, 1987) by 'Third World' development discourses on the value of expert knowledge and trickle-down approaches to change. Generally, these approaches have been and continue to be framed by views and agents from industrialized countries; consequently, they have focused on economic growth and suffered from a poor understanding of rural reality. In fact, there has been and continues to be little dialogue with those for whom the development programmes were designed; communication structures still have a vertical, top-down dynamic in practice (Flick, 2008). This vision of development translates into top-down action in which the poor are the objects of first-world benefactors' administrations and become seemingly dependent on these external inputs and visions to develop (McMichael, 2010). Thus, problems, solutions and the direction of 'progress' were all defined by experts who were disconnected from local reality (Freire, 1979; Chambers, 1987; Ashby et al., 2000; Pokorny and Johnson 2008a, 2008b) whereas the views of the smallholders were often overlooked (Ingles et al., 1999; Gibson et al., 2000). Consequently, many of the development programmes failed and often continue to fail to adequately correspond to the reality of the smallholders. In other words, the livelihood systems and capacities of smallholders often do not match the aims, objectives and inputs of development initiatives, and vice versa.

The ripples of the past in today's frontier dynamic

The construction of the first roads connecting the Amazon region to the remaining South American countries resulted in a massive flood of new actors into the region. In Peru, the first road entering the Amazon was previously constructed in 1945, running from Lima, the capital, to Pucallpa (854 km). However, heavy rainfalls remain a problem today, regularly eroding the highway and frequently even cutting it off from use. In Ecuador, the first road to the Amazon reached the city of Puyo in 1947. Later, in 1957, this road was further extended to Tena. Brazil only started road construction in the 1960s with the highway from Brasilia to Belém (1,985 km) and then completed the Transamazon Highway in the 1970s. The Bolivian Amazon was first connected to much nearer Brazil; terrestrial transport from the urban centres of La Paz (919 km) and Santa Cruz de la Sierra (1,996 km) to the Amazonian city of Riberalta today still function only occasionally during the dry periods, so air transport is necessary between these cities.

Since then, in addition to the thousands of families who arrived during numerous colonization projects or those who had migrated to the region on their own in search of a better life, there has also been an influx of more capitalized players trying to seize opportunities in the Amazon. The state frequently encouraged these actors through the provision of incentives in the form of subsidies and tax breaks, the generous transfer of land tenure rights or, more indirectly, the construction of roads. In view of the tremendous costs associated with systematic exploitation of the region, governments have also continuously attempted to attract international investors by allowing access to the resources of interest in exchange for infrastructural investments. These processes were often (and still are) catalyzed by corruption (Parker et al., 2004, Thomas, 2008). NGOs, the majority of them with an environmental agenda, have also entered the region in great force.

The entrance of these national and international actors, each with specific interests in the resources of the region, inevitably resulted in a confrontation with the vulnerable local populations living in the aforementioned post-colonial conditions characterized by informality, weak social organization and limited adaptive capacities. This collision with recently arrived actors in new and dramatically changing institutional contexts generated opportunities to improve the quality of life in the region, both for the local families who had lived there for decades – in some cases for centuries – in conditions of poverty and dependence and for thousands of immigrants trying to escape poverty in their areas of origin (Medina et al., 2008). However, despite the emerging demographic shift, the historical structural inequalities – with a limited number of powerful actors controlling a huge number of powerless families – continued to significantly limit the possibilities for poor families to take advantage of the emerging opportunities. In agreement with the findings of Wienold (2006), the findings from our study areas suggest that 'agrarian capitalism grew in the peripheries, not by fruitful investments and the generation of employment but through the application of extensive production models based on the repression and exploitation of small producers and natural resources'.

Land grabbing

In the Amazon, as in the majority of agrarian economies, access to land is the fundamental cornerstone of production. Moreover, land ownership often translates into political clout for the owner (Bardhan, 2000). In this way, those actors who have ready access to capital are naturally interested in ensuring their right to property and access to the land and the properties' natural resources. In line with this concept, we found in all study areas that powerful families had successfully managed to accumulate and continued accumulating large tracts of public land and land used by smallholders. In these acts of 'land grabbing', they found effective ways to ensure legal access to the resources of their interest. As highlighted by Barreto et al. (2008), these actors systematically use their power, their capacity to take advantage of the flaws in the legal land registration system and their ability to navigate within that system.

We found the most impressive examples of land grabbing in the study areas in Brazil. There, the business of 'organizing' legal ownership of land has its own term, 'grilagem', and it is performed by semi-professional persons called 'grileiros'. As described by Benatti et al. (2008), their job is to provide legal land documents that hide the origin and legal status of the land in question by initiating a parallel process of new formal recognition of the ownership or user rights by the state. The generated land titles or certificates of authorization of use are then sold to interested parties such as cattle ranchers or loggers. Estimates indicate that there are close to 100 million hectares of illegally occupied land in Brazil, of which approximately 85% are in the Amazonian states of Amazonas and Pará (IPAM, 2006). However, based on our field-collected data, we find these estimates to be rather conservative. For example, in the municipality of São Felix do Xingu in *Pará*, which covers three active frontiers, the land registered in the notary offices exceeded the size of the municipality by 10 million hectares. Obviously, to perform land grabbing at such a level, it is necessary to have the means for physical control of the land, good personal contacts with the notary offices responsible for land registration and financial resources to pay for informal services.

Therefore, land grabbing is almost exclusively undertaken by capital-endowed actors. This observation is confirmed by the experiences emerging in the course of a survey by INCRA (National Institute of Settlement and Agriculture Reform), which, in December 1999, notified all alleged owners of land in Brazil holding more than 10,000 hectares and required them to present their land titles. In Pará, only 207 of the 73,000 landowners did not respond. However, this handful of alleged owners (0.28% of the total number of registered properties) had titles in their name that represented 34% of the total registered area in the state of Pará (Di Sabatto, 2001). Similarly, a 2005 survey conducted by ITERPA (the land institute of the state of Pará) in conjunction with the state notary found more than 9,000 irregular documents of land ownership among large farms registered in the names of individuals or entities such as logging and agribusiness companies claiming (and blocking other uses for) approximately 500 million hectares of land in the state. The fact that the total area of the state is only 125 million hectares may indicate the extent of fraudulent registration and land grabbing (Greenpeace, 2005).

In the majority of study areas, governments have attempted to attract large farmers from outside the Amazon to invest in production activities by establishing public financing programmes, offering subsidies or enacting attractive tax laws (Wehrlich, 1978; Kohlhepp, 1991; Ryder and Lawrence, 2000; Klein, 2011; Giugale et al., 2003). Simultaneously, the continuous expansion of the road system and the provision of utilities such as energy and other facilities have gradually improved the production conditions in the region. In response, companies and other capitalized actors have started to more systematically buy land to benefit from the improved conditions, particularly in those areas where new roads or the paving of dirt roads were planned. Margulis (2003) named this phenomenon 'frontier speculation'; driven by the expectation of increasing land prices, such speculation differs from 'frontier consolidation', which describes investments in land intended to generate profits based on productive use. More recently, the production conditions in many areas became so attractive that agricultural expansion no longer depended on public subsidies. Instead, the profit expectations from the production of soybean, oil palm, timber and so on for export markets increasingly became the principal motive behind the occupation of land in the Amazon (Pokorny and Montero, 2007). However, the success of external capital-endowed actors in occupying vast areas was still facilitated by weaknesses in the governments and the low prices of land and labour (Rodrigues, 2004; Sobral Escada et al., 2005; Reimberg, 2009; Ribot and Peluso, 2003; Welch, 2009; Sikor and Stahl, 2011).

A different form of land occupation, observed especially in Ecuador and Peru, was the purchase of large tracts of intact tropical forest ecosystems for biodiversity conservation. The capital for these often significant investments originated principally from environmental foundations, but individuals with sufficient capital were also involved. In several cases, the forest lands were effectively protected against use and invasions in addition to being managed for ecotourism; in other cases, especially in more remote settings, the forests remained without oversight. Although these environmental protection efforts have recently gained increasing attention under the label 'green grabbing' (Fairhead et al., 2012), the effect of such processes in the study areas were marginal in comparison with the overall dynamic, largely because effective protection proved to be rather challenging after the purchased land became accessible due to infrastructure expansion.

The case of cattle ranching

The production of meat has greatly increased and now holds major importance in the region, especially in Brazil and Bolivia. In Brazil alone, more than 10 million head of cattle were slaughtered in 2007, a 43% increase since 2004. Moreover, one-third of the total meat exported by Brazil originated in the Amazon (Smeraldi and May, 2008). The study performed by Godar (2009) along the Transamazon Highway revealed that ranching in the region was closely associated with the process of land accumulation in the hands of largeholder families. Despite the fact that, when the Transamazon colonization programme started in the 1970s, the majority of the land was distributed to small producers, by 2007, less than one-third of the settled area was still in their hands (Figure 4.1). Almost half of the occupied area had become controlled by large producers, specifically cattle ranchers, and another 22% had become occupied by medium-sized producers with properties between 200 and 600 hectares. Figure 4.1 also shows that many largeholders were located far from the road, where it was possible to practise extensive ranching and where the prices of land were relatively low. This finding indicates that there were two types of cattle ranchers active in the Transamazon region, each following different strategies: first, those who preferred to occupy areas farther away from the main road, utilizing the lack of state presence and control mechanisms to bolster production; and second, those who were more interested in land that was well connected to infrastructure and who were thus highly interested in buying land from smallholders, often resorting to violence to obtain the land that they preferred (Godar et al., 2012a).

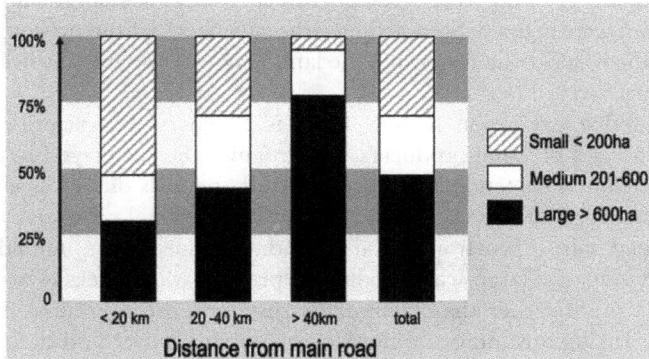

Figure 4.1 Proportion of area occupied in 2007 by actors with different property sizes in the municipalities of Brasil Novo, Anapú and Pacajá along the Transamazon Highway in relation to distances between their properties and the main road (adapted from Godar et al., 2008)

The data presented in Table 4.1 confirm this process of smallholders' expulsion from their original lots by large producers in the studied area, illustrating the close relation between the size and growth of individual properties. Almost 90% of the large landholders increased their property area compared with only 16% of the smallholders, and the degree of expansion was much greater for the largeholders. On average, the large producers increased their properties by 1,380 hectares, whereas the smallholders expanded their lots by only approximately ten hectares. These strong and successful efforts by largeholders to take possession of land were also confirmed by Rodrigues (2004), who found that more than 80% of all property titles in public areas approved between 1992 and 1998 by the governmental agency responsible for the Brazilian Amazon (INCRA) were for areas of more than 1,000 hectares.

In all study areas of the four project countries, ForLive observed processes of land accumulation similar to those that occurred along the Transamazon Highway, although to different degrees; often, this land accumulation reflected

Table 4.1 Tendency for growth in individual property area among landowners with different-sized properties in the Transamazon Highway region (n=94 interviews)

Actor category	N	Farmers with increased individual property area (%)	Average increase in land area (ha)
Smallholders (with less than 100 ha)	70	16	10
Medium-sized producers (200–600 ha)	14	50	98
Large cattle ranchers (more than 600 ha)	10	90	1,380

historical processes. In the state of Acre, for example, almost all the areas along the main road were actually occupied by large ranchers, but the presence of fruit trees along the roadside indicates that the land here was also initially occupied by small farmers.

We found that expansion to new areas is a general characteristic of the land-use strategies of large producers, in particular those engaged in extensive cattle production. This tendency to expand land area was due to unsustainable land-use practices that were poorly adapted to the regional ecosystems, causing an accelerated rate of environmental degradation due to the immediate and short-term vision of a large-scale production process, which seeks to accumulate benefits and profit rather than optimizing sustained use over time (Godar et al., 2012b). Taking advantage of the lack of public control and their political power, many large producers systematically abused the opportunity to exploit new areas and extensively amplify their production area. In view of the land-use dynamics observed in the study areas, it can be concluded that the production systems unfolded differently at different scales. Specifically, whereas the capitalized actors invested their profits in the purchase of new areas, small-holders generally tended to reinvest in the same plot to intensify or optimize their production system.

Other examples of acquisition of control over resources

In contrast to the study areas in Bolivia and Brazil, where the land-use dynamics were strongly influenced by cattle ranchers, in the Peruvian Amazon in particular, agroindustrial companies cultivating palm oil (*Elaeis guineensis*) for the production of biodiesel were another dominant largeholder group (März, 2008; Blum, 2009). According to Jonasse (2009), many companies, spurred by a government that created favourable conditions for international investors, began cultivating palm oil, mainly in regions with adequate infrastructure such as around the cities of Pucallpa and near Puerto Maldonado, where the Inter-Oceanic Highway was under rapid development. In the Ecuadorian Amazon, where the majority of areas were still dominated by small settlers and indigenous groups (Giugale et al., 2003), the first indications of expanding agroindustry arose as a consequence of large investments in the construction and paving of major roads primarily in the northern regions (Killeen, 2007). However, the prevailing strategy of the majority of the companies was mainly to integrate the small landholders in the production chain as producers of raw materials.

In Ecuador, as in many Amazonian countries, prominent examples of the appropriation of resources by external actors were related to the exploitation of minerals, petroleum and gas, mainly by multinational companies (Acosta and Falconi, 2005). In general, after profitable stocks were found, the families living in the affected areas had little opportunity to influence the ensuing dynamic (Thomas, 2008) and, in many cases, they were forced out of their settlements. The same dynamic was observed with regard to the many water dam projects initiated for hydropower generation (Sanchez, 2007). Due to the relatively flat relief typical of large parts of the region, the areas flooded by the dams were often immense, and flooding thus affected a large number of (often indigenous) families who were resettled to new villages where they did not have the possibility to continue their traditional economic activities. The dams that are planned or currently under construction in the Amazon are estimated to affect an area of more than 35,000 km² (Holt-Giménez and Spang, 2005; Spang, 2005; Earthjustice, 2007).

In the study area located in northern Bolivia, we found a specific type of land occupation originating from the historical phenomenon of the 'barracas', which were large areas of forest controlled by a few powerful families with the right to harvest Brazil nuts. Historically, 'barracas' were established for rubber exploitation, which continued despite low world market prices until the 1980s. Some families had managed to build veritable empires, exercising a strong influence in regional and national politics. However, these families were not able to impede a number of land reforms in which many of them formally lost their land tenure. For example, the locally famous Hecker family was left with only approximately 1,500 ha of legally acknowledged private property out of the more than 300,000 ha initially demanded. Nevertheless, confronted with this new situation, many of these families looked for other ways to retain control of the forests by utilizing their political influence. Thus, several families were able to receive legal NTFP concessions for large areas, whereas others, such as the Hecker family, converted their areas into peasant communities, which they continued to informally control by taking advantage of their monopolistic position in the commercialization of forest products. Thus, despite various efforts at agrarian reform, the powerful families still continued to control huge areas of forest as well as the indigenous people (Stoian, 2005).

Forest concessions can also be interpreted as a method for capital-endowed actors to ensure control over valuable resources. In the forest concession system, the government sells the right to exploit timber from public forests based on formally approved forest management plans. Despite the enormous difficulties inherent in ensuring the quality of management by the concessionaires (Gray, 2002), this concession approach was utilized in all four project countries to make public forests available to the private sector. Some companies were successful in obtaining concessions to control large areas over a cutting cycle of approximately 20–40 years. Considering the limited productivity of timber in primary forests, a medium-sized timber company needs, at minimum, approximately 30,000 hectares to supply a saw mill during this period. In the study areas, however, due to the complex bureaucratic processes of bidding and the necessity of large investments in heavy machinery and equipment, it was almost impossible for smallholders to access these concessions and fulfil the established legal requirements. Even the participation of medium-sized logging companies, politically promoted by several governments (e.g. Brazil's National Press, 2005), faced great difficulties due to a lack of capital and the insufficient professional capacity of the vast majority of timber companies active in the region to meet stringent legal requirements (see Plate 13). Thus, independently of several attempts to protect the local population and ensure the continuation of their traditional forest uses, families living inside or around the concession areas were generally excluded, often by law, from the use of the forests exploited by the concessionaire (Sabogal et al., 2008a; Carvalheiro et al., 2008; Ibarra et al., 2008; Martínez Montaño, 2008).

In several cases, traditional or indigenous communities obtained the right to utilize large areas of forest. For example, in the Ambê initiative in the Tapajós National Park located in the state of Pará, Brazil, local river communities, strongly subsidized by the government and NGOs, managed approximately 25,000 hectares of public forests (Medina and Pokorny, 2008, 2011). There were also a number of indigenous and extractivist reserves on which the right to exploit the forest in accordance with authorized forest management plans was granted exclusively to the local families. We found this model in multiple places: in Bolivia, there were public forest concessions with municipal government support called Agrupaciones Sociales del Lugar (ASL); in Peru and Ecuador, there were indigenous reservations; and in Brazil, there were the Sustainable Development Projects (Projetos de Desenvolvimento Sustentável; PDS) established by the government. Peru and Bolivia also provided concessions for the use of NTFPs such as Brazil nuts. However, in the majority of these initiatives, smallholders generally faced severe difficulties in meeting the technical and administrative requirements set by law for harvesting the products from their forests (Medina and Pokorny, 2008, 2011; Benneker, 2008; Pacheco et al., 2010). Interestingly, in view of these difficulties, the private sector started to systematically explore possibilities for accessing the forests surrounding these communities (Lima at al., 2003; Pantoja, 2008; Masias, 2011).

Exploitation through integration

In general, we found an increasing tendency for large producers of meat, soybeans, timber and biodiesel in the study areas to apply strategies for the integration of smallholders into their production chain as producers of the raw material needed to supply their industries; these strategies thus served as an alternative to pursuing private property. Instead of appropriating the smallholders' land, the large producers preferred to indirectly determine and control what smallholders produce by utilizing their privileged position in the markets and distribution networks, which the smallholders were less informed about and had less ability to access. For the smallholders, these partnerships occasionally offered one of the few possibilities for generating a reasonable income, whereas the large producers had the advantage of leaving a large part of the responsibility and economic risk with the families, who took over part of the investment (with a smaller cushion) through their contributions of labour and land. In this manner, the large producers achieved more flexibility (and less risk) in expanding or reducing their business in response to market dynamics, reserving their resources for investment in processing capacities, logistics or other operational sectors rather than tying up capital in land acquisition. This strategy was especially observed in areas characterized by emerging efforts for land reform and high prices for fertile land with reasonable access to infrastructure, and in places where smallholder families had achieved good local organization, generally with the support of development organizations (Pokorny et al., in press b; Masias, 2011).

In the timber sector, these relationships between the larger companies and smallholders historically had (and still have) great importance. In fact, much of the timber used by the Amazonian timber industry originates from small properties (Nepstad et al., 2004; Asner et al., 2005; Lentini et al., 2005; Brandão Jr. and Souza, 2006). Additionally, in the study areas, a large part of the timber processed by the saw mills was harvested on the properties of the smallholders, who sold their timber at generally low prices to the timber industry or traders in the form of logs, pre-processed planks or even as standing trees. Generally, the companies harvested the timber from the smallholders' forests without paying attention to currently existing technical regulations and without asking the permission of the families (Medina and Shanley, 2004). In the study areas, this informal harvest of timber by loggers was still the most common method of fulfilling the timber industries' demand for raw material (Medina et al., 2008; Sato et al., in preparation). With improving legal and institutional forestry frameworks (see Chapter 5), the number of management plans required to legally harvest the timber from the smallholders' forests increased dramatically. These management plans, however, were developed by timber companies, which established 'partnerships' with the communities and took responsibility for the 'legalization' of harvesting operations because the process was overly complex and beyond the smallholders' technical and legal expertise (Lima et al., 2003). In Bolivia, for example, in one of the management plans formally approved by the governmental authorities, more than one million hectares of land were inhabited by traditional and indigenous communities, with this pattern representing an increasing trend (Pacheco et al., 2010). Considering that large forested areas were recently designated for smallholders and communities (White and Martin, 2002; Sunderlin et al., 2008), it is probable that smallholders will remain the major suppliers of raw timber materials in the region, whereas large companies will continue to control processing and logistics and to encourage relationships based on dependence.

We found several other examples in which smallholders were linked to commercial processes under the control of external actors. As described above, a handful of powerful families formed paternalistic relationships with indigenous people to harvest Brazil nuts and, in this way, built controlled empires known as the 'barraca' system for Brazil nuts in northern Bolivia (Stoian, 2005). Even after several agrarian reforms, these powerful families were able to ensure that forests with the greatest economic potential for Brazil nut harvesting remained at their disposal. Thus, many of the descendants of these workers, after working in the 'barracas' in a form of semi-slavery, found themselves left only with the forest lands of low productivity or those located in remote areas after the land reforms. To secure their livelihood basis, they were often forced to hire themselves out as daily workers during the harvesting period for Brazil nuts. This situation thus reinforced the historical relationship of dependence between the workers and their 'patrons' (Pacheco, 1992; De Jong et al., 2006).

Another example of the various ways in which capital-endowed actors continue to exploit smallholders was observed in Pará, where large cattle ranchers in older frontier areas began to expand their production areas by systematically using smallholders to appropriate the remaining available public land. The ranchers leased cattle to small colonists in settlements originally granted to smallholders, who were mainly engaged in forestry production ('Projetos de Desenvolvimento Sustentável'; PDS). By accepting these cattle from the large ranchers, the smallholders were spurred to convert their forests into pastures. In doing so, they assumed the responsibility and risk of this illegal activity. Once the forests were fully converted, the families were expected to leave their properties so that the ranchers were able to buy the now-deforested lot for little money. Thus, the Brazilian settlement policy initially directed towards smallholders indirectly promoted the accumulation of land in the hands of large cattle ranchers (Godar, 2009).

In the Brazilian state of Pará, we also found that capitalized actors sometimes even established types of private colonization programmes. Without authorization, they constructed roads in areas of public forests, sold small plots of land along these roads to migrants and then took advantage of these areas and the cheap labour force to supply their own industries. For example, they bought timber at low prices from the forests and also bought various agricultural products, especially rice, grown in the recently cleared areas. Often, as observed in the region near the city of Santarém, the large producers still benefited in a third way, as they repurchased the area formerly sold to the smallholders at a low price after it was deforested, using it for the large-scale production of rice and, more recently, soybeans in other regions. In this way, many capitalized actors continued to benefit from large tracts of land, with little risk and large profits. As in the case of the cattle ranchers, the investors followed a strategy of gradual expansion so that they continuously increased their property (Benatti et al., 2008).

In Peru, especially along the larger roads to the cities of Pucallpa and Puerto Maldonado, the biofuel industry, in addition to purchasing land for the production of palm oil, systematically explored the possibility of production on smallholders' land. By offering loans and technical assistance, often also supported by the state, they managed to quickly convince a rapidly growing number of small producers to replace their traditional production systems with the cultivation of palm oil (*Elaeis guineensis*). Because of the attractive prices and the existence of promotional programmes, the rural producers showed great interest in such an option, both in the Peruvian study area and in other places (Ferreyros and Medina, in preparation; März, 2008).

In the study area in Ecuador, in contrast to the study areas in the other countries, the economic elites were not as omnipresent in the Amazon region, mainly because major investments in road construction had only recently started, and colonization of the Amazon in Ecuador and land reforms were much more consistently targeted at small families (Giugale et al., 2003). Other important factors leading to the reduced presence of largeholders in Ecuador include the absence of a distinct dry season, which hinders the use of fire in the upper part of the Amazon, and a gently rolling topography that discourages the use of heavy machinery. However, after recent investments in the construction of roads, large producers eagerly invaded the Ecuadorian Amazon. Of note, we observed, as did Killeen (2007), a vigorously increasing dynamic in the timber and biodiesel industries along the new and improved roads.

In addition to the various ways of using smallholders for the appropriation of land and resources, we frequently found examples of the capital-endowed actors using smallholders for their own purposes in more direct ways. For example, a large number of the municipalities and district governments in the region were led and administered by large cattle ranchers and loggers who were elected and re-elected due to their capacity to finance campaigns and invest in corruption (Salgado and Kaimowitz, 2003). Often, these powerful actors were easily able to mobilize families against initiatives that would have harmed their interests. In many situations, they organized and even paid families to participate in demonstrations against reforms or initiatives for greater social democratic control. This mobilization is especially prevalent in the timber sector in Brazil, where the large saw mill owners repeatedly organized protests as a strategy to pressure the legal authorities to scale back their efforts to control illegal harvesting and authorize management plans that had been halted for technical or formal deficiencies more quickly. In their analysis of these social dynamics in the municipality of São Felix de Xingu, Rede Geoma (2004) concluded that many times, those smallholders and land workers economically and socially dependent on the large landholders were extremely vulnerable to political instrumentalisation.

Cultural deterioration and environmental degradation as a consequence of smallholder marginalization

In all the study areas, we observed strong investments in the expansion and improvement of roads. These efforts were generally accompanied by the attempt of governmental authorities to formalize the institutional context, establish mechanisms of control and law enforcement and improve public services. In general, road construction opened up opportunities for the smallholders to improve their often-precarious situation by allowing better access to markets and important public services such as education and healthcare. This process of social inclusion in remote Amazonian areas also provided a great opportunity for these families to have their civil rights acknowledged and thus break out of the historically unjust and exploitative relationships with local patrons and other capitalized elites. Access to markets opened new opportunities for commercialization and consumption, generating income while satisfying essential and urgent demands. In addition, for many of the migrants and colonists coming to the region as part of this dynamic, the advancing frontier offered, for the first time in their lives, the chance to establish a solid base for a livelihood that could sustain their family by managing their own resources and their own land.

However, the parallel arrival of actors with more experience, more resources, better information and greater technical knowledge naturally provoked new conflicts over the resources of interest. Typically, the better position and where-withal of capital-endowed actors within the production chains resulted in an increased pressure on local actors, who were generally less skilled and less prepared to seize emerging market opportunities (Pokorny et al., 2012). In the same vein, the growing presence of the state generally implied various restrictions and a confrontation with formal institutions and bureaucracy. The necessity of capital to invest in competitive production systems systematically favoured the more capitalized producers.

As a consequence, in the study areas, the new contexts left smallholders facing an unfavourable situation in which they were often unable to maintain control over their resources and their property (Godar, 2009; Blum, 2009). The transformation of large public areas into private land instigated a process of dislocation for many families – traditional, indigenous, settler and migrant families – who, in search of a better life, began to migrate to urban centres or settled on public land without the availability of even the minimum economic, social and institutional conditions needed to establish a solid foundation for life. Instrumentalized by capital-endowed actors, these families did not participate in the dynamic growth of the region as independent producers but as 'products of technical and institutional modernisation and agricultural expansion' without a significant share of the benefits (Becker, 2005). In the words of Wienold (2006), they simply 'appeared and disappeared as "autonomous" entities along a continuum of changing incentives for migration and colonisation, however always maintaining their precarious working and living conditions'.

Effects on production systems

In all the study areas, the smallholders attempted to adapt their socio-productive systems to take advantage of the opportunities generated by emerging markets. This adaptation implied drastic changes including in particular an intensified and often simultaneous expansion of production that often ignored ecosystem limitations. With improving linkage to international markets, a number of new products were also introduced – a process generally accompanied by an increasing degree of specialization and the application of new technologies that often required higher capital input compared with traditional land-use practices (see Chapter 2). An increasing dependence on external inputs and technologies was a common consequence of this process.

In several of the case studies, emerging possibilities for access to attractive markets for forest products, timber and some NTFPs induced a massive intensification of harvesting, which contributed to forest degradation and, in several cases, even resulted in the exhaustion of marketable forest resources in the short term. In other case studies, in particular those in which the families were more oriented towards agricultural markets, the intensification of agricultural production and the application of modern technologies triggered an accelerated transformation of primary forests into intensive agricultural cropping systems, particularly for crops that require a relatively high level of soil fertility, such as rice. Thus, this process resulted in a gradual disappearance of the spatio-temporal micro-mosaic of production components typical of many traditional production systems (see Chapter 3). In fact, the analysis of the case studies revealed that even some of the smallholders' productive activities that were generally perceived as being an intrinsic part of their traditional systems were actually adopted from external models. Examples include tree plantations, the commercial harvest of forest products and cattle ranching (Figure 4.2). Production components still perceived as traditional were extraction of forest products for subsistence, shifting culti-vation agriculture, single-tree cultivation, home gardens and livestock production.

Undoubtedly, the product that generated the strongest environmental and social impacts in the region is cattle. In the study areas, cattle were typically kept on rather large areas, and this land use was often gradually expanded to and beyond the limits of the individual properties. More recently, some smallholders also began to cultivate plants on a larger scale for the production of biofuels in intensified systems generally financed and technically supported by companies interested in purchasing the products. In particular, we found that smallholders in study areas with fertile riparian zones tended to spontaneously replace their traditionally cultivated crops with those that temporarily reach high prices in national or international markets. One example was the substitution of traditional banana cultivation for papaya in the region of Pucallpa. However, several studies (e.g. Morley, 1995; Homma, 2006) demonstrate that these 'boom' phases often

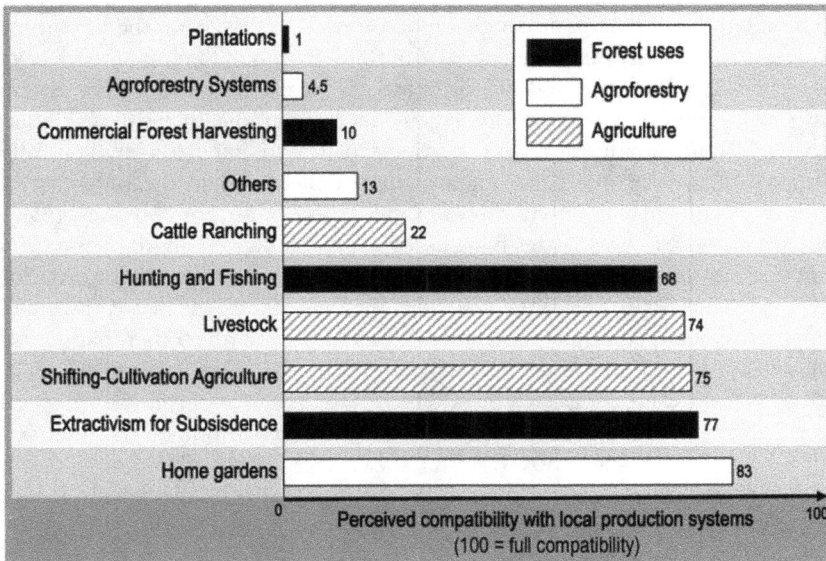

Figure 4.2   Evaluation of the level of compatibility between external models of land use and traditional systems of production in the view of producers in Ecuador, Peru and Bolivia

do not last for reasons such as market collapses due to local oversupply or pests, phenomena which are significantly accelerated by the massive expansion and intensification of single-product cultivation during a boom phase.

Cultural reorientation

The studies conducted in ForLive show that smallholders' efforts to adapt to the requirements of emerging markets not only branched into significant lifestyle changes in terms of technical considerations, but also caused significant changes in the social organization of the smallholders. The assimilation of market logic, dominated by the idea of competitiveness, and entry into the monetary sphere often initiated a process of individualization and a verticalization of the local social system that deeply affected the smallholders' way of life (Porro et al., 2008; Campos, 2009). Naturally, these changes were more dramatic in indigenous groups and traditional communities because the original pattern of social organization was very different (Porro et al., 2008; Gasche and Vela, 2012). As a result of this process of 'social modernization', combined with the increased participation of families in the formal education system and better access to media, including television, we observed in all case studies a process of gradual replacement of traditional local cultures and values with those of the 'modern, globalized world'. Available local knowledge and traditional skills and practices accrued over generations became less important and, in several cases, disappeared completely, as also observed by Shanley and Rosa (2004).

This process of cultural reorientation occasionally caused large problems for the smallholders. This was strikingly demonstrated by a study on the importance of medicinal plants for the indigenous groups of Shipibo-Conibo, in the Ucayali region of Peru (Lange, 2008). Over years of increased contact with external actors, the importance of modern medicine increased radically for these families; concurrently, the appreciation and knowledge of traditional medicine diminished drastically. However, due to strong structural deficiencies in the public health sector in Peru, typical of many countries in the region, the communities suffered from insufficient access to the benefits of the promulgated modern medicine. Consequently, the indigenous groups encountered an even more precarious health situation than before because, on the one hand, the newly promoted modern medicine was not accessible to them but, on the other hand, their ability to apply local traditional medicine had been devalued, and sometimes the knowledge was even completely lost. The development implications are clearly described in the reflections of an indigenous producer in one of the interviews with a ForLive researcher (Box 4.1). In view of these implications, an increasing number of experts suggest promoting the great potential of medicinal plants in health systems in rural areas (Alexiades, 1999).

We observed another example of the implications of this process of cultural reorientation in the case study of the indigenous community of Callería in Peru. This community participated for several years in a 'Community Forestry' project that resulted in the FSC certification of their operation (Campos, 2009; Nalvarte,

2011). The traditional dynamic of activities in the community was mainly oriented towards satisfying the basic daily needs of the family. After fulfilling this obligation, the families usually devoted their time to social activities or hunting and fishing. As observed by Henkemans (2001), this menu of traditional activities was not subject to a fixed schedule or dates for completing certain tasks. This rhythm of life changed dramatically when the 'Community Forestry' project was introduced, as it implemented completely new practices necessary to ensure the proper performance of the forest operations in accordance with the management plan and economic demands. The project introduced a new way of working including task sharing, preparation of plans, control activities, a regular working rhythm that spanned eight hours of the day, time intervals, days off such as holidays and Sundays and the payment of fixed salaries. This change in the very conception of the rhythm of life implied profound changes in social organization and in internal relationships within the community. In particular, these people may have strengthened their managerial capacities and significantly increased their self-confidence, but they also had much less time to devote to family, neighbours and other social activities (see also Chapter 6).

Box 4.1 *Reflections of a Shuar man (Ecuador) about the changes in his community (interview performed in the year 2007, translated from Spanish)*

'There are some plants and animals that should not be eaten, some of them, in accordance with our ancestors, are spirits, dirty spirits, for example "la lechuza", "el venado", which should not be eaten because they said they are the devil ..., but today we are following some traditions and customs, but not all, because some have already forgotten their ancestors and are not giving any more importance to this knowledge. Now, what the sons are saying is "ah ... these beliefs are only myths, they have no value"; this is what the young people say ....

[These beliefs] our ancestors had, this is the wisdom of our ancestors, and now there are infinite things that are being really forgotten. For the gardens, there were songs that my wife and I do not know anymore because our grandparents died before we grew. But when I was young, my mother could have managed to teach me a few songs. However, as I am not a woman, I was not allowed to learn these songs because the women have other songs for going to the garden ..., for going to the field and when to start weeding, then they sing, there is a spirit which they say is the "Nunguí" ....

Civilization was a total failure, although we learned to read and write, we learned to thank to the missionaries. Although a bit evil, doubtlessly we learned to speak Spanish, but this is not our language. Between us we speak our language; at least we must keep our language.'

Hindering development with traditional models

In the ForLive study regions, we found that the relationships between smallholders and a number of diverse groups of capital-endowed actors operating in the region were mediated by various mechanisms of power, which allowed the most privileged groups to define the way in which smallholders used their resources (Medina et al., 2009a, 2009b). Often, several large producers managed to oblige the smallholders to adopt natural resource management schemes in accordance with their interests using mechanisms of direct and indirect coercion including sabotage, threats and blackmail. In all the study areas where large producers had access to the resources of interest through the smallholders, we found inequitable relationships with a high degree of dependence. Although the relationships with the large landholders and small companies had a more paternalistic character, the larger companies and multinationals used more impersonal and contractual mechanisms to ensure their interests. This latent imposition of external models and the paternalistic relationships with external actors systematically limited the possibilities for smallholders to develop their own ideas on the use of natural resources and organize around common interests for the implementation of their ideas.

This effect was most apparent in the analysis of discourses related to the issue of forests and smallholders: we noted an increasing influence of external actors on the organization of the local families (Medina et al., 2009a, 2009b; De Koning, 2011). Timber companies argued that, unlike the local communities, they have the technical capacity to ensure the professional and sustainable management of Amazonian forests, whereas environmental agencies highlighted the fact that market influences and the illegal deals with timber companies led communities to over-exploit their resources in exchange for very limited and short-term financial benefits. In light of this situation, they argued that communities should be trained to manage their forests according to models defined by experts and specialists. We became aware that it was very difficult for smallholder families to develop and articulate their own positions in this argument and, consequently, many families simply adopted the prevailing discourses and the actions associated with them. Interestingly, in the interviews, we found that the producers who collaborated directly with timber companies adopted the discourse utilized by the timber sector, whereas families who participated in 'Community Forestry' projects copied the discourses used by environmentalists and the supporting NGOs. Apparently, the 'copycat' discourse on sustainable forest management was limiting the possibilities for the smallholders to not only develop their own ideas of how to manage their resources, but also to have their existing ideas legitimated by society. It also became clear that the often-paternalistic relationship and proximity to external actors had the potential to bias local organizations and provoke conflicts with those members of the community who were not involved in these external-based relationships. Consequently, many families became disconnected from their own representative organizations.

In the study of initiatives that were supported by NGOs and governmental development agencies, it became obvious that, in their attempts to implement models and ideas for development and sustainable resource management defined by experts and external organizations, these organizations widely ignored the local institutions and norms developed by the families to regulate access to and manage their resources (Pokorny and Johnson, 2008a, 2008b). In these cases, we found that very little space had been created to allow the communities to participate more actively in the design of norms and regulations that better reflect their needs and realities. However, a sequence of studies on the regulatory frameworks for forest management by smallholders in Bolivia, Ecuador, Peru and Brazil, showed that compliance with the legally defined forestry norms imposed exaggerated transaction costs for the communities who wanted to legally harvest their forest products and that the implementation and enforcement of these regulatory frameworks constituted another way in which external actors exercised power mechanisms that made life difficult for the smallholders (Sabogal et al., 2008a; Carvalheiro et al., 2008; Ibarra et al., 2008; Martínez Montaño, 2008). According to the existing legal frameworks for forest management, smallholder communities had to obey the rules designed to achieve sustainable management of their forests. To this end, the state agencies exercised control and punished the violation of these regulations. However, it became increasingly difficult for communities to fulfil these rules because they were designed principally for and by large-scale forest management operations such as timber companies (see also Chapter 5).

Smallholder marginalization

The findings of ForLive clearly confirm that the rural societies in the Amazon region have faced an intense process of transformation. In their attempts to utilize new opportunities on the frontier, local families sought to adapt to market requirements, which were generally determined by external actors. Thus, the development process observed in the study regions demonstrated a significant weakening of local knowledge, management schemes and social organization, leading to increasing homogenization of cultures and local specificities as well as the 'disappearance' of valuable differences in perspectives regarding how to both meet the needs of the rural poor and provide environmental conservation; in particular, perspectives arising from the smallholders themselves were silenced. Consequently, despite the often significantly improved well-being of smallholder families, our analysis revealed a simultaneous process of cultural and environmental homogenization that contributed to an accelerated process of ecosystem transformation and, consequently, its degradation.

To some extent, this dynamic shows certain similarities with the history of peasantry in Europe, albeit within a very different social and environmental context. In Europe, the small rural farmers were integrated economically, socially and culturally into a more globalized society (Jollivet and Mendras, 1971; Jollivet, 1974). With agricultural modernization, the European smallholders lost their relative autonomy while increasing their dependence on industrial inputs to ensure the competitiveness of their agricultural production. Consequently, the farmers increasingly involved themselves with large-scale markets oriented towards the demands of urban consumers. However, the undeniable success of the productivist model in agriculture eventually suffered its own visible crises, particularly in the 1980s, as manifested in several phenomena such as overproduction, the marginalization of the sector, declining profits and, as a consequence of the above, the exodus of small farmers (Lamarche, 1993). In view of the very different contextual conditions in the Amazon, it remains an open question whether and to what extent such a process will be repeated or how it might be avoided.

Plate 1   Amazonian smallholders represent a huge variety of ethnicities and cultures resulting from many indigenous groups and several waves of colonization: (A) shipibo in Peru (L. Hoch); (B) 'wild' settlers ('caboclos') in Brazil (B. Pokorny); (C) family immigrated from the Andean region in Peru (L. Hoch); (D) colonist from South Brazil (G. Medina); (E) descendants from rubber tappers in Bolivia(L. Hoch); and (F) descendants from Africans brought to Amazonia ('quilombola') in Brazil (G. Medina)

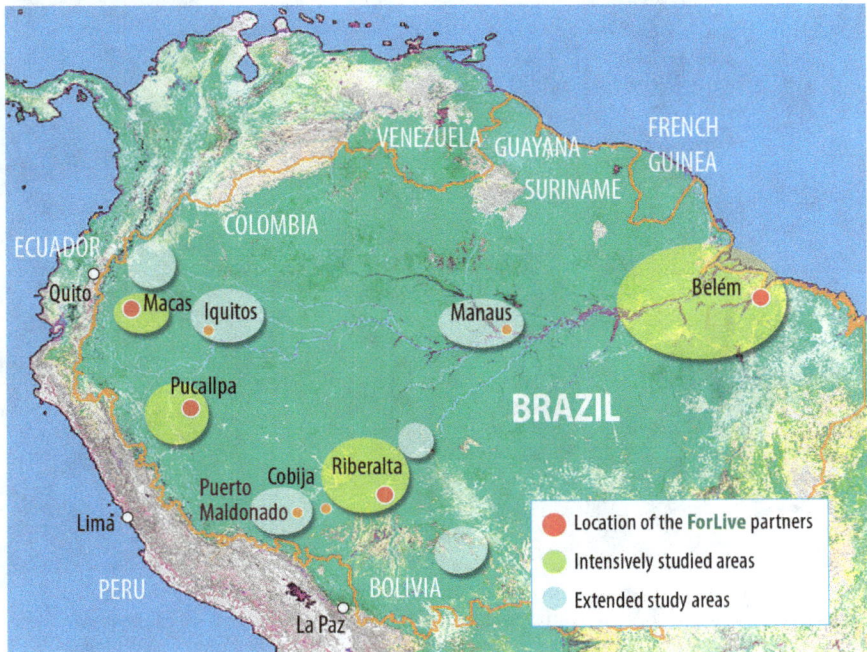

Plate 2   Study areas of the ForLive project

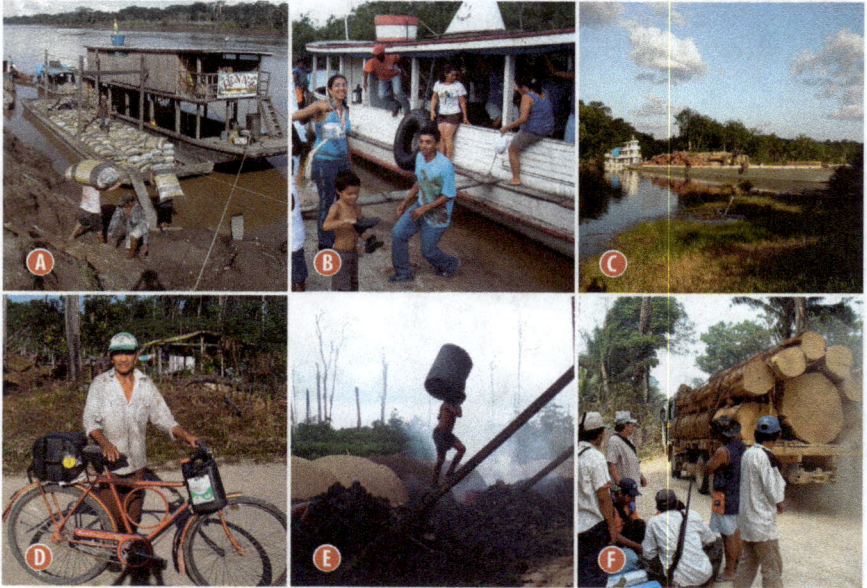

Plate 3 In many areas, rivers continue to be the principal mode of transportation for (A) production (C. Quette), (B) people (G. Medina) and (C) timber (J. Johnson), while in the advancing agricultural frontiers, roads are the principle mode of transportation. (D) Poor people using bicycles and motorbikes for transport (B. Pokorny), however heavier loads such as (E) charcoal (B. Pokorny) and (F) timber (G. Medina) require trucks.

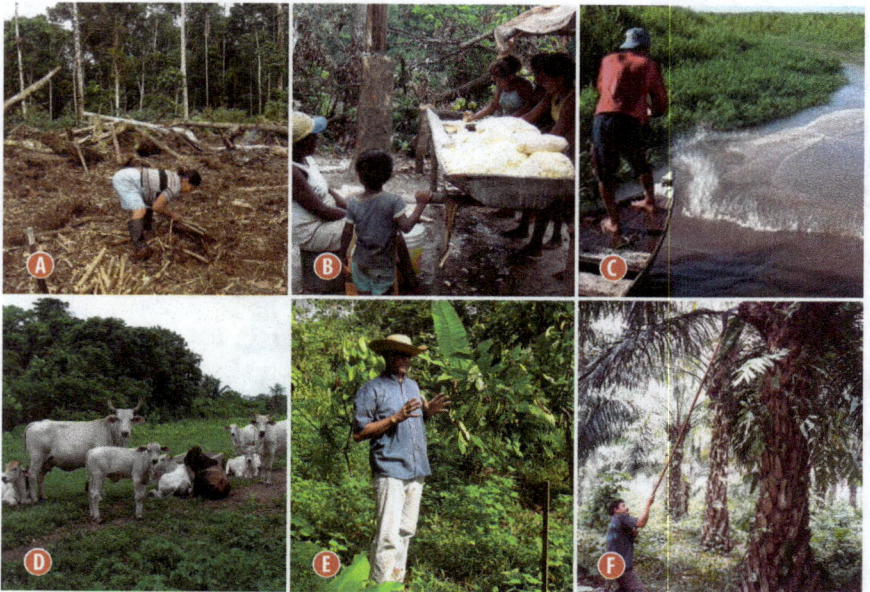

Plate 4 Amazonian smallholders are engaged in a wide variety of agricultural activities. Typical traditional activities are the production of cassava (*Manihot esculenta*) (A) harvesting (J. Johnson), (B) processing (B. Pokorny) as well as (C) fishing (ProVarzea); more recent land uses include (D) cattle (L. Hoch), (E) perennials such as cocoa (*Theobroma cacao*) (L. Hoch) and (F) oil palm (*Elaeis guineensis*) (L. Hoch)

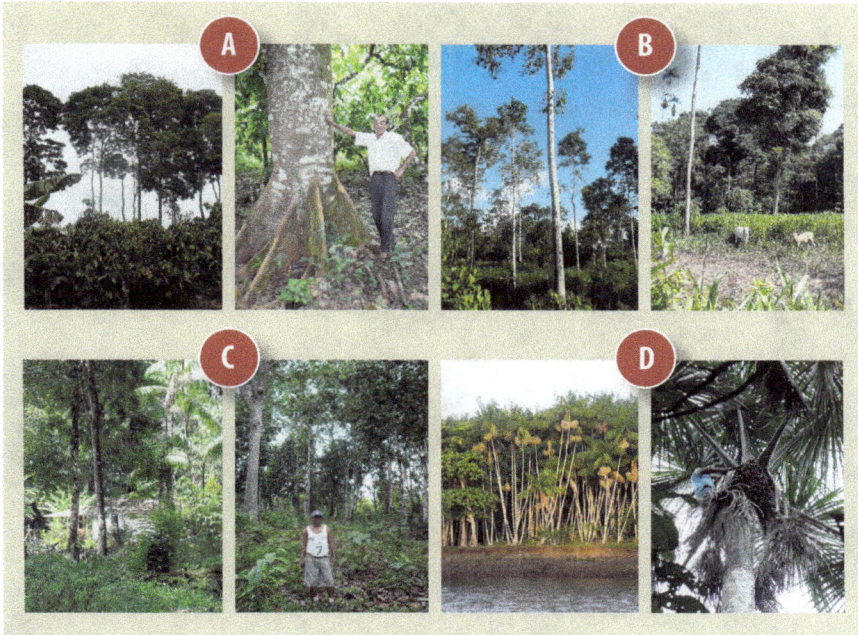

Plate 5   Examples of local forest management practices: (A) single-tree planting of mahogany (*Swietenia macrophylla*) in a cocoa (*Theobroma cacao*) plantation; (B) promoting regeneration of late pioneer species (e.g. *Pollalesta discolour*) in pastures in Ecuador and Brazil; (C) fruit production in home gardens; and (D) intensification of natural açaí (*Euterpe oleracea*) stands (photos by L. Hoch)

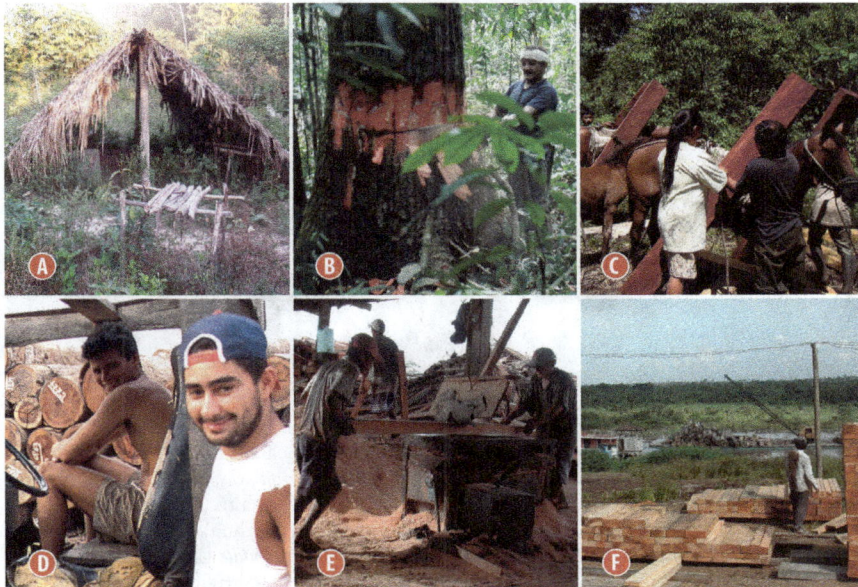

Plate 6   Many smallholders are engaged in the informal timber sector: (A) forest camp (B. Pokorny); (B) technically deficient cutting of timber (R. Muehlsiegl); (C) transporting pre-processed boards with horses (L. Hoch); (D) local transport 'entrepreneurs' in Brazil (R. Muehlsiegl); (E) local saw mill (B. Pokorny); and (F) commercialization along the rivers (B. Pokorny)

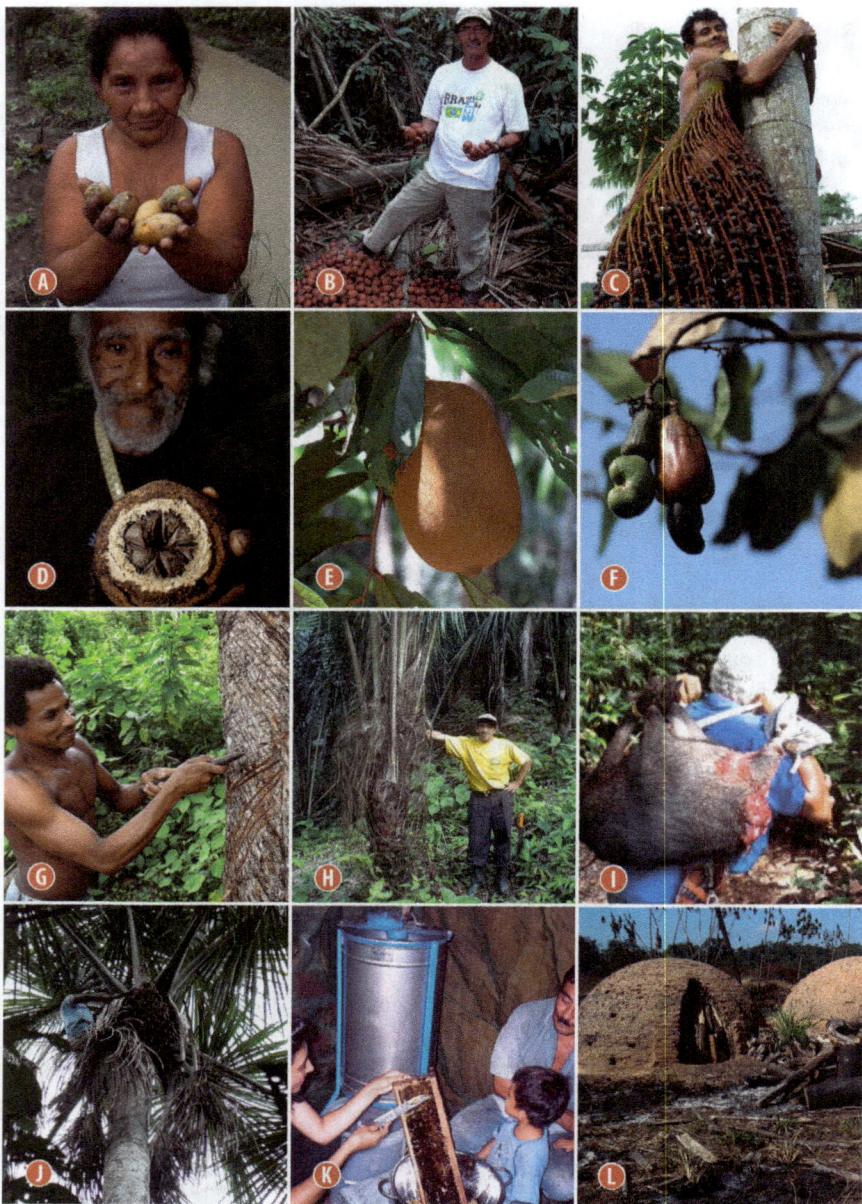

Plate 7   For many indigenous and traditional families, the use of the wide range of non-timber forest products for subsidence and local commercialization is an important livelihood component: (A) uxi (*Endopleura uchi*) (G. Medina); (B) moriche palm (*Mauritia flexuosa*) (L. Hoch); (C) turu palm (*Oenocarpus bacaba*) (S. Ferreira); (D) Brazil nut (*Bertholletia excelsa*) (G. Medina); (E) rubber (*Hevea brasiliensis*) (G. Medina); (F) cashew (*Anacardium occidentale* L.) (R. Muehlsiegl); (G) cupuazu (*Theobroma grandiflorum*) (L. Hoch); (H) palm fiber (*Aphandra natalia*) (L. Hoch); (I) game (G. Medina); (J) batauá palm (*Oenocarpus bataua*) (L. Hoch); (K) honey (J. Johnson); and (L) charcoal (B. Pokorny)

Plate 8   The entire region is experiencing a strong urbanization dynamic: (A)
           traditional river house (B. Pokorny); (B) traditional river village (L. Hoch); (C)
           rapidly growing cities (Pucallpa in Peru) (B. Pokorny); (D) rural electrification
           remains often limited to the urban centres (L. Hoch); and (E) typical
           'timber' city (Tailândia in North-eastern Brazil) (R. Muehlsiegl), Amazonian
           metropolis (Belém, capital of Para, Brazil) (R. Muehlsiegl)

Plate 9   In advancing frontiers, people are rapidly changing the landscapes: (A)
           area prepared for shifting cultivation in a forested landscape (G. Medina);
           (B) structured landscape created by small colonists (L. Hoch); (C) large cattle
           ranchers compete with smallholders on land (R. Muehlsiegl); (D) along the
           roads, forests are completely transformed in other land uses (R. Muehlsiegl);
           (E) yearly burning is expected to maintain the quality of pasture (R. Muehlsiegl);
           and (F) large areas are already free from trees and forests (L. Hoch)

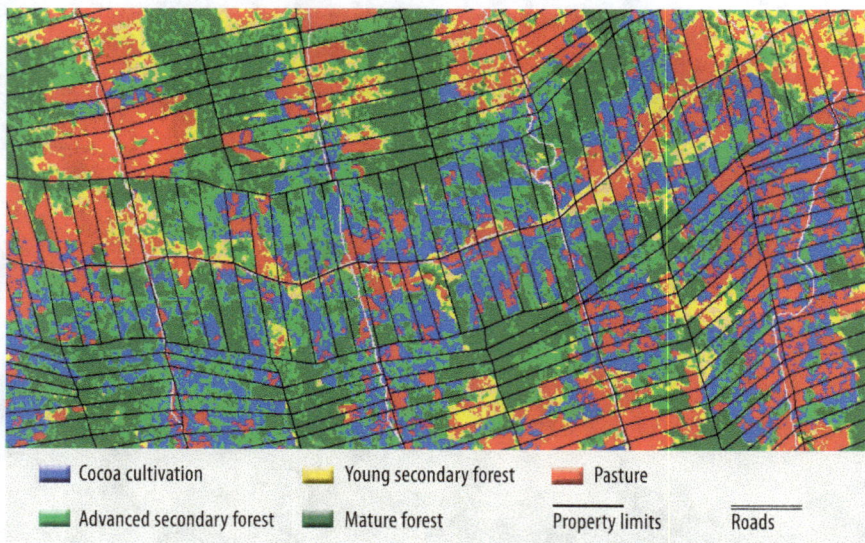

▣ Cocoa cultivation	▣ Young secondary forest	▣ Pasture
▣ Advanced secondary forest	▣ Mature forest	Property limits	Roads

Plate 10   Example of the landscape diversity created by smallholders (Medicilândia, Brazil) (adapted from Godar et al., 2008)

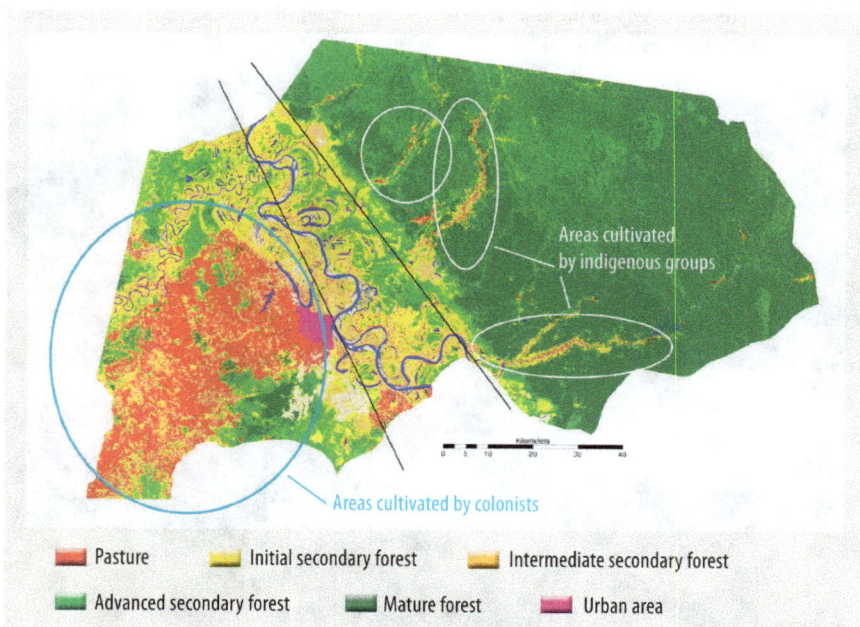

Areas cultivated by indigenous groups

Areas cultivated by colonists

▣ Pasture	▣ Initial secondary forest	▣ Intermediate secondary forest
▣ Advanced secondary forest	▣ Mature forest	▣ Urban area

Plate 11   Region of Pucallpa (Peru) with areas dominated by indigenous people (right) and by colonists (left) (adapted from Godar et al., 2008)

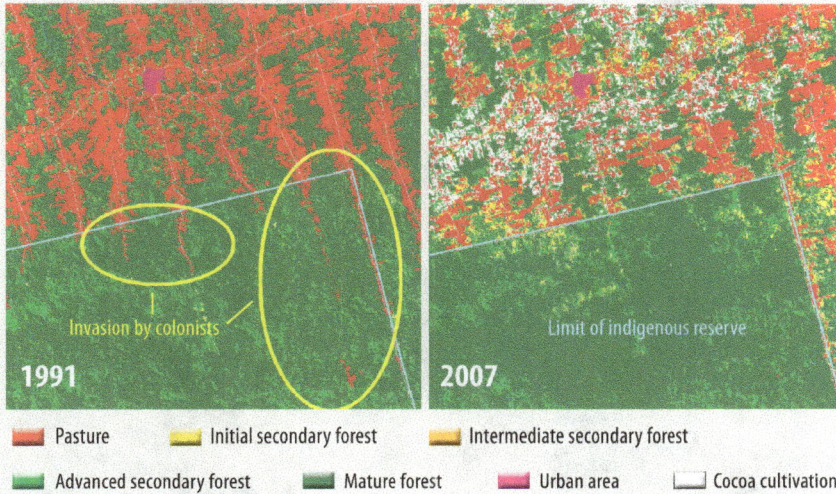

Plate 12   Satellite images of part of the municipality of Medicilândia located along the Transamazon Highway in 1991 (left) and 2007 (right) (adapted from Godar, 2009)

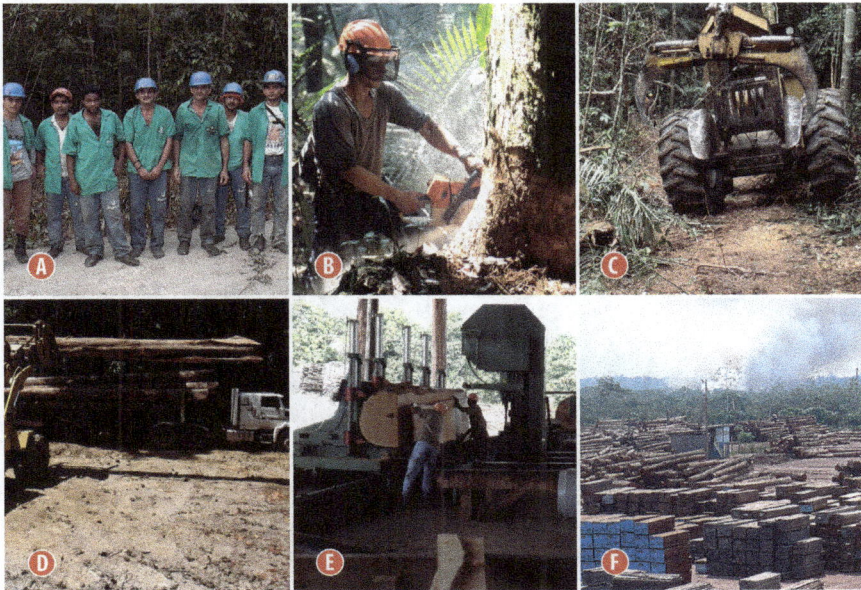

Plate 13   The legal approach for large scale timber management by far exceeds smallholders' capacities: (A) Systematic tree inventories as central element of Reduced Impact Logging require well qualified teams (J. Souza); (B) tree cutting requires fast-working professionals (R. Muehlsiegl); (C) the use of skidders is only meaningful in large areas over long time periods (R. Muehlsiegl); (D) the purchase of specialized machinery for transport is incompatible with local realities (R. Muehlsiegl); (E) saw mills require professional management to be profitable (R. Muehlsiegl); and (F) mechanized timber harvest allows for attractive profits, but also requires large investments (R. Muehlsiegl)

Plate 14 Community forestry projects implemented by NGOs and governmental agencies with international funds can be found throughout the region: (A) sign indicating a community forestry project (G. Medina); (B) awareness raising and capacity building are typical NGO activities (J. Johnson); (C) promoted technologies often lack compatibility with the smallholders' reality (J. Godar); (D) portable sawmills are expected to increase smallholders' share of profits, but technical difficulties often arise (G. Medina); (E) transporting the timber out of the forest remains a challenge (G. Medina); and (F) despite significant volumes harvested, the profit margin is quite limited due to high production costs (G. Medina)

Plate 15 Transportation conditions are difficult in the Amazon, especially during the rainy season: (A) transport on rivers can be time consuming (B. Pokorny); (B) the majority of roads are constructed by private actors such as timber companies (B. Pokorny); (C) bituminized roads allow fast transport but require high maintenance costs (D. Martins); (D) many villages are cut off in the rainy season (L. Hoch); (E) heavy rain falls and high temperature make the roads highly sensitive to damages (L. Hoch); and (F) rivers represent a serious obstacle for long distance transport (C. Quette)

5 The challenge of supporting smallholders

In all the study areas, the development frontier was advancing, although at different speeds depending on specific local factors. The way in which the development played out generally did not meet the proposed expectations. A general increase in wealth observed in and around the emerging urban centres was juxtaposed with only moderate benefits to families living in the rural areas, in which social unrest and environmental degradation greatly impacted the traditional livelihood basis (Pokorny et al, in press b). In view of this imbalance, governmental and non-governmental organizations began to reformulate ideas regarding ways in which the local people themselves could more effectively benefit from emerging opportunities. In many of the case studies, we found initiatives resulting from these considerations (see Table 1.5). This chapter presents findings from the analysis of these cases.

The numerous development initiatives targeting the rural poor managed to improve the precarious situation of many families (see also Box 6.1), but they have not managed to alter the general process of social differentiation creating increasing social gaps between rich and poor (UNDP, 2010). In fact, our findings from the case studies and study areas even indicate a latent tendency for the majority of the observed development initiatives to accelerate the prevailing development dynamic, with both the positive and negative associated consequences. Thus, instead of tackling the historically embedded and highly unjust institutional frameworks and strengthening the immanent capacities of smallholders, the majority of development initiatives have simply attempted to adapt the socio-productive schemes of the locals to the requirements of markets or organizations' agendas. By widely ignoring the potential of locally developed systems of social organization and production, this approach subtly contributed to a process of cultural homogenization that contributes to unjust social structures and accelerates deforestation and environmental degradation (Pokorny et al., 2012).

Although there is a large number of development initiatives, with different strategies, tools and actors, we identified several common characteristics. For instance, nearly all initiatives in the study areas were intended to generate so-called win-win situations. Thus, they sought to improve the well-being of the local families – mainly through the generation of income opportunities – while also expecting to contribute to environmental conservation. This approach was most apparent in those development projects that targeted a more effective use of natural resources through profound modernization of local socio-productive systems and their integration into financially attractive markets. To achieve this aim, the supporting organizations in the case studies provided training, gave advice, purchased materials and equipment, made credit and funds available and provided payments to the families. In addition, in all project countries, there were massive efforts aimed at improving regional planning and territorial control, reforms of the legal and institutional framework regulating the management of forests and other natural resources and varying degrees of decentralization and investment in mechanisms for environmental governance. The following section summarizes the observations from the study areas related to these themes.

Development organizations seeking win-win situations

The effective integration of the Amazon region into national and global economies has been and is still understood to be a crucial precondition for the development of the region and the improvement of the often precarious situation of the rural poor (see, for example, the Infrastructure Development Programme for the Integration of the Region, IIRSA: see http://www.iirsa. org/index01.asp?CodIdioma=ESP; IDB, 2006; Brazilian Ministry of Finance, 2008, 2009). Since the 1980s, the governments of the Amazonian countries have massively encouraged private and corporate investment in the region using fiscal and financial incentives (Binswanger, 1991), including the expansion of large-scale extensive cattle ranching (Barclay et al., 1991; Hecht and Cockburn, 1989; Pacheco, 2005), soybean production (Nepstad et al., 2006) and extractive activities such as timber harvesting, mining, the exploitation of oil and gas and the construction of hydroelectric dams (Bunker, 1985). More recently, the private sector has made a massive investment of its own capital to access resources of interest. Governments have supported this market-based approach, arguing that private capital and professional know-how are needed to exploit the region's resources (Pokorny et al., 2012).

Since the Rio Summit in 1992 in particular, Sustainable Forest Management (SFM) has emerged as a means of reconciling the twin development goals of generating economic growth and conserving natural forest ecosystems (UNCED, 1992). In parallel, poverty reduction goals, social movements and NGO initiatives have worked to support local peoples' rights (Schmink and Wood, 1992; Chapin, 2004) and, since the 1990s, environmental organizations have started to highlight the importance of the local poor as crucial contributors to the successful environmental protection of Amazonian forests (Schmink and Wood, 1992; Nepstad and Schwartzman, 1992; Becker, 2005). Consequently, many governments and development organizations started to tackle the question of how to adapt the SFM approach to local forest managers as a means of addressing rural poverty. Throughout the Amazon, development organizations and national governments started to promote the marketing of products from locally managed forests (timber and non-timber forest products, NTFPs) to generate much-needed income for smallholders and maintain the provision of environmental goods and services (Homma, 2006; Almeida et al., 2006; Hoch et al., 2009).

The application of such integrated development approaches to natural resource management was intended to link poor families more effectively to markets to ensure that the value added to forest products would provide an incentive to reduce deforestation (Pokorny and Johnson, 2008a, 2008b). In search of attractive prices, the majority of these initiatives oriented production towards external, often international markets. Many initiatives also considered the possibility of adding value by eliminating intermediaries or involving locals in the processing of the harvested products (Amaral and Amaral Neto, 2005).

Another common aspect of almost all governmental and non-governmental initiatives for local development analyzed in the study areas was the often-implicit understanding that smallholders had to change their productive systems to increase effectiveness and achieve compatibility with the demands of globalized market chains (Pokorny et al., 2012). Obviously, the governments and development organizations inherently considered the profound modernization of smallholders' production systems as a crucial prerequisite for taking advantage of market opportunities and generating urgently needed income. This understanding involved not only the technical aspects of production, but also social organization. Thus, the modernization of the entire socio-productive system became an integral component of the win-win approach of integrated development projects, which implies two important assumptions: first, the understanding that smallholders are not effectively managing their resources; and second, that the way in which smallholders use natural resources greatly contributes to environmental degradation (Sayer, 1995; Wunder, 2001). However, this widely circulated stereotype of the 'vicious circles' was not confirmed by our studies (see also Chapter 3).

The unspoken predisposition of development organizations to dramatically intervene in the smallholders' systems of production and to adapt them to the 'modern world' necessarily implies a certain devaluation of the wide range of cultures, knowledge and capacities of the local populations (Pokorny et al., in press b). The interventions observed in the case studies generally aimed to implement packages designed by specialists that were based on scientific knowledge of production with business models oriented towards specialists' ideas about optimal recipes for local development (Pokorny and Johnson, 2008a, 2008b). This approach, which is focused on the transfer of externally defined management schemes, remained rooted in the philosophy of agricultural modernization from the 'Green Revolution' developed in the 1960s and 70s, and it was found to run through all the policy sectors in the study areas.

For example, in the health sector, we observed package interventions to introduce modern medicine in indigenous contexts that displaced traditional medicine (see Chapter 4). In the agricultural sector, the majority of the packages offered by governments imposed the introduction of new products for commodity markets and the application of advanced technologies that relied on significant capital input and mechanization. In the forestry sector, the development organizations fostered the management of natural forests in accordance with the guidelines for Reduced Impact Logging (RIL) developed by forestry experts primarily for the management of large forest areas by companies, which necessarily implied certain incompatibilities with local cultures (see Chapter 2). In other contexts, governmental agencies and NGOs encouraged the producers to establish plantations with seedlings produced in nurseries. We found only very few development projects that attempted to help continue the traditional ways of utilizing forest products or to optimize the systems of agricultural production based on actual practices discovered and used by the smallholders themselves.

Generally, the NGOs in all study areas played a huge role in defining and implementing local development initiatives and creating several norms within the area that were defined by this external influence. This dynamic was interpreted as a direct consequence of two general tendencies: the weak interest of the national and municipality governments in supporting smallholders and the intensification of international organizations' engagement in the region, particularly following the Rio Summit. Seeking opportunities for effective implementation of their programmes, the international donors generally disregarded the governmental agencies as ineffective counterparts and instead promoted NGOs with well-qualified and motivated staff as implementers of their programmes. Consequently, national and international NGOs, many of them with a strong environmental agenda, gained crucial importance in the rural sector of the region. In several parts of the study areas, these alliances between international organizations and local NGOs effectively assumed the role of a temporary government in term of their technical assistance and extension functions. Only more recently has the public sector begun to revitalize its efforts to address the needs of the smallholders through (rather rare) credit programmes (Pokorny and Johnson, 2008b). As a consequence of the business models of international organizations and their finance routines, the NGOs conducted and still conduct the majority of their activities within temporarily financed projects with objectives and modi operandi set by the donors. In general, these projects invest relatively significant human and financial resources over relatively short time periods and in relatively small areas, with very specific objectives.

Integrated development projects

The development initiatives observed within the framework of ForLive sought to achieve situations of mutual benefit in the hope of improving the lives of the smallholders while simultaneously contributing to environmental conservation. As an immediate prerequisite for these types of initiatives, the development organizations assumed that there is a large commercial potential for forest use based on sustainable management practices that can contribute significantly to improving the lives of rural families. This concept of the win-win situation was implemented in so-called integrated projects by many programmes and projects in the forestry sector, including initiatives for 'Community Forestry', smallholder plantations and agroforestry projects (see Plate 14).

In another common outlet, the projects fostering 'Community Forestry' hoped to optimize the use of primary forests by smallholder families by generating financial income and thus increasing the value of their resources and their interest in conserving them (Sabogal et al., 2008b). There were several major initiatives in the region that followed this model. One was named the sub-programme ProManejo, part of the Pilot Programme for the Conservation of the Amazon (PPG-7), which supported 52 promising initiatives in the Brazilian Amazon. In Bolivia, within the framework of forestry legislation reform, various programmes, such as the BOLFOR II project, supported communities in the legal harvest of their forest products by assisting them in complying with the administrative and technical requirements of the newly defined regulations and proposed RIL technologies. In this vein, programmes in Peru and Ecuador were also implemented to promote forest management by smallholders. All the case studies involved in 'Community Forestry' received strong support from NGOs, often lasting from several years up to a decade. Interestingly, in view of the barriers and incompatibilities generated by the legal regulations and processes within which these projects functioned, the NGOs often became lobbyists for a simplification and adaptation of the legal framework. However, they seldom managed to overcome the inflexibility immanent in the laws and regulations (Pokorny and Johnson, 2008a).

We also identified many programmes and projects that promoted tree plantations for smallholders (Hoch et al., 2009). These initiatives often attempted to combine the production of marketable forest products for income generation with the aim of reclaiming degraded land. The Brazilian government, for example, created the National Silvicultural Plan for Native Species and Agroforestry Systems (PENSAF), with the goal of distributing 12 million seedlings annually (MMA et al., 2007). Similarly, the state government of Pará inaugurated the programme 'A billion trees', intended to encourage 120,000 families to plant forest species in their plots (Government of the State of Pará, 2008). In Peru, the National Reforestation Plan, initiated in 2005, attempted to establish almost 900,000 ha of commercial tree plantations and to demarcate areas of environmental protection while encouraging the participation of smallholders (MINAG and INRENA, 2006). In the Ecuadorian Amazon, the National Reforestation Plan (PNFR) aimed to encourage family farmers to establish more than 100,000 ha of 'social plantations' and agroforestry systems (MAE, 2006b).

In all the countries studied, there were also extension programmes for agroforestry that were meant to strengthen diversified local smallholder production (Hoch et al., 2009). The majority of the proposed agroforestry systems sought to enable the simultaneous production of agricultural crops and tree products in relatively complex production systems in order to use the ecological potential of the sites effectively while simultaneously ensuring or even increasing ecological stability, especially by maintaining soil fertility (see, for example, Milz, 1997). In northern Bolivia, we found that the first NGO initiatives promoted the diversification of production through agroforestry systems as early as the 1990s, with several initiatives that had been active for more than 20 years. In Peru and Bolivia, the programmes aiming to provide alternatives to the cultivation of coca leaf were essential for the support and execution of agroforestry projects on a small scale. In addition to the various initiatives led by NGOs and international projects, for the past several decades, governmental programmes have also participated in the promotion of agroforestry systems. Examples of these programmes include the sub-programme for the establishment of demonstration areas (PDA/ PPG-7) and ProAmbiente, both in Brazil (Almeida et al., 2006; Chapin, 2004; Simmons et al., 2002; UNDP, 1997).

Support strategies

In order to enable smallholders to cope with the technical, managerial and financial requirements of the promoted 'packages' and to ensure competitiveness in the envisioned markets, supporting organizations provided training, advice, material, equipment and often also funding (Pokorny and Johnson, 2008a). In essence, we found in the case studies that development organizsations employed three types of support strategies: (1) 'Pilot projects', often related to 'Community Forestry', to implement technical-organizational packages in cooperation with selected communities, with specific families within a community or even with single producers. The establishment of demonstration areas was expected to provide a visual understanding of the viability of the promoted models as a basis for further dissemination. Generally, the involved organizations invested considerable effort and resources over relatively short time periods (generally between two and five years) to establish the pilot sites (Medina and Pokorny, 2008, 2011); (2) Intensive long-term working support was intended to create a long-standing relationship between the supporting organization and the collaborating families to support the implementation of the technical-organizational package through the establishment, management, harvesting and commercialization phases. The few examples we found in the region were mostly related to agroforestry initiatives, or they evolved from ongoing cooperation between development organizations and locals in the follow-up phase of 'Community Forestry' pilot projects (Hoch, 2009; Hoch et al., 2009); (3) Extensive initial support, observed principally in public credit programmes or, more frequently, programmes promoting tree plantations. These programmes, found in all study regions, generally intended to establish nurseries for the production of plants – preferably mahogany (*Swietenia macrophylla*) and cedar (*Cedrela odorata*) for timber or fruits such as oranges, lemons (*Citrus sp.*), cocoa (*Theobroma cacao*), peaches, tembe (*Bactris gasipaes*) and cupuazu (*Theobroma grandiflorum*) (Montero, 2007). Seedlings were then distributed to the families free of charge or at highly subsidized rates. These programmes also provided technical assistance and training. They occasionally offered financial incentives during the establishment phase, after which support was gradually stopped (Hoch, 2009; Hoch et al., 2009). Moreover, as a consequence of increased awareness of the global challenge of climate change, but also to make Sustainable Forest

Management more financially attractive (Putz et al., 2008), the international community instigated a new generation of market-based approaches for environmental protection, the so-called 'Payments for Environmental Services' (PES; Wunder, 2005; Hall, 2008; Pokorny et al., 2012).

The most visible form of action by development organizations in the study regions was the implementation of 'pilot projects'. In many places, NGOs implemented technical-organizational packages developed by specialists at the level of an individual, family, group or community to demonstrate viability, to learn about the eventual possibilities for improvement, to visually show the function of the package and to contribute to its propagation. Implicit in this idea was the assumption that, once initial support was provided and after a consolidation period, the producers would be able and willing to continue to apply the practices and guidelines of the package on their own. However, contrary to this assumption, our analysis revealed that the costs of pilot projects were significant for development organizations and there was minimal long-term success. Generally the 'pilot project', following the logic of temporary injections into the system, provided significant resources for a short period of time – approximately three to five years on average – to achieve the results expected by the donor within the timelines set by the project. The investments included diverse activities such as consultancies, capacity building and technical assistance, in addition to payments for equipment, machines and logistics. As an example, Figure 5.1 demonstrates the value of the investments in numerous pilot initiatives for 'Community Forestry' in the Brazilian Amazon (Medina and Pokorny, 2008, 2011).

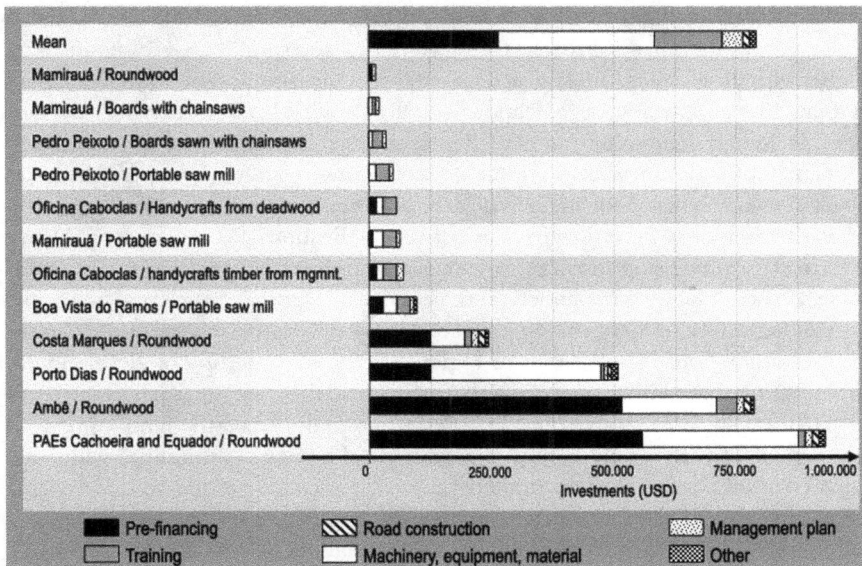

Figure 5.1   Initial investments in the establishment of pilot projects for 'Community Forestry' in the Brazilian Amazon (adapted from Medina and Pokorny, 2008, 2011)

Depending on the scale of the operations, the technology used and the degree of vertical integration in the analyzed case studies, development organizations invested between US$20,000 and US$800,000 on equipment, machinery and capacity building to establish the pilot project. Thus, the investments were usually high and greatly surpassed the financial capacity of the participating families. Although for larger initiatives, the costs for heavy machinery and pre-financing harvesting operations were most significant, in smaller initiatives, the costs for the technical assistance required for compliance with legal requirements repre-sented the most important outlay. In all the initiatives, the costs for training were considerable, varying between US$10,000 in the smaller initiatives of less than 20 m³ of harvested timber per year up to US$40,000 for initiatives incorporating processing and manufacturing. These numbers were confirmed for similar initia-tives in Bolivia, Peru and Ecuador, albeit with slightly lower personnel costs due to lower salary levels (Medina et al., 2009c; Vos et al., 2009; Robles, in preparation). The analysis also revealed significant investments made by the participating families – not considered in the values presented above – including time invested in meetings and training courses and the dedication of large areas taken out of their own land and production cycles, subjecting the families to land management restrictions and foregone income opportunities without compensation (Villacis, 2010).

Even in plantation programmes, where support was mostly limited to the distribution of seedlings and extensive technical assistance during the initial planting phases, Hoch et al. (2012), like other studies (Current and Scherr, 1995; Sunderlin et al., 2005), calculated costs of US$500–3,500 per ha. Costs in the few programs with more continuous support (longer than a few years) and a wider and more universal reach, i.e. more than a few families, were significantly higher. In these cases, mostly associated with the promotion of agroforestry systems, the supporting organizations usually provided more intensive support, often including the establishment of demonstration sites and training centres and sequences of training courses and field visits during relatively long periods. In an attempt to compensate for the absence of local markets, the supporting organiza-tions themselves became involved in commercialization and invested significant resources in processing facilities, organization and logistics. Often, the continu-ation of such initiatives depended on the continuous receipt of external financing.

Although development initiatives were prominent in the discourse of govern-ments, donors and NGOs, even in regions with a high density of development organizations, good infrastructure and proximity to urban zones (such as the areas near Riberalta, Bolivia and Pucallpa, Peru), the majority of the families had no direct contact with development organizations. The few communities with such contact usually received only limited support, typically catering to one specific issue such as a certain system of production, specific health-related support or temporary support to improve the education situation. Moreover, these forms of support were generally defined according to the priorities of external organiza-tions (Depzinski, 2007; Pokorny and Johnson, 2008b; Biedenweg, 2009). Keeping in mind the high costs associated with the pilot project approach, or even more pointedly, an intensive long-term accompaniment approach, and given the large number of families, many of which lived in extremely remote contexts, the probability that a rural producer would be selected as a partner in such initiatives was rather low. In the forestry sector, Medina et al. (2009c) estimated that only 2% of the rural families in the study areas had contact with a development project. We also found that, for pragmatic reasons, the organizations often tended to select the same local partners repeatedly (Depzinski, 2007), often resulting in an accumulation or a sequence of different development initiatives with a relatively small group of families.

As a result of the increasing influence of development organizations that include a social component, and in view of the only moderate and somewhat ambivalent outcomes of efforts so far (see Chapter 6), a (minor) part of the available financing was invested into small-scale projects with a more active involvement of smallholders. To date, however, this type of support for locally driven initiatives remains limited in size and scope and seriously lacks the human and financial resources to be effective at a larger scale (Pokorny and Johnson, 2008a, 2008b). In the study areas, the most functional possibility for smallholders to receive technical assistance and public rural extension services was ultimately through existing credit programmes (Miranda, 1990; Pokorny and Johnson, 2008b). These credit programmes, however, were seldom compatible with the interests and capacities of the smallholders and, due to the formal and bureaucratic requirements, many families were not able to access them (Pokorny et al., in press b). This dynamic highlights another general problem, namely, that many of the agencies and agents involved in services for technical assistance and rural extension were often inappropriate for meeting the needs of local families (Pokorny and Johnson, 2008b). Many of these technicians never received adequate training for communication and facilitation and had little or no experience and knowledge of participatory processes for collaboration. Thus, they naturally tended to concentrate their assistance efforts on technical and financial issues (Montero, 2007). This dynamic helped exacerbate an 'expert-driven' system in which the smallholders themselves were excluded from being active, thinking participants and instead followed 'expert' advice for the duration of short-term projects and did not want to or were unable to troubleshoot the technical problems of the new systems in the long term – eventually leading to the failure of the project (Flick, 2008). This situation was even further aggravated by the fact that many technicians were commonly selected by the government according to political and personal criteria and not necessarily because of their qualifications. Consequently, technicians working at the local level often tended to represent political commitments and not local interests.

National support measures

In addition to the development initiatives that directly interacted with the smallholders and focused on implementing externally defined technologies and organizational models, during the last decade in nearly all Amazonian countries, there have been large investments made to improve general living conditions and provide more effective management and protection of natural resources, particularly forests. Correspondingly, there was a push to include local people more in the development and decision-making mechanisms. More specifically, countries adjusted their legal and institutional frameworks by including new forestry laws and regulations; they performed large-scale ecological-economic zoning of land tracts; they demarcated protected areas; they prepared National Forest Plans; and they invested to varying degrees in the decentralization of environmental management. These processes have widened the bases for decision making, contributed to dialogue and also had positive effects on the formalization of smallholders' rights to access land and natural resources. However, because the efforts concentrated on environmental conservation objectives, there has been limited impact on wider policy issues or in other sectors of the economy. Despite such reforms, the legal and institutional frameworks for the use of natural resources by smallholders still suffer from major incompatibilities with local realities (Pokorny et al, in press b).

Efforts for regional planning and control

In all the countries studied, we found that the governments – often supported by international collaborations – invested in regional planning as a strategic approach to improve their control over the actors who were present in or moving into the still-forested regions in the Amazon (Sombroek and Carvalho de Souza, 2002; Pitman et al., 2007; Larson and Soto, 2008). The most prominent initiatives were the National Forest Programmes and the initiatives for the Economic-Ecological Zoning, which included the demarcation of significant proportions of protected areas, and the Tarapoto Process (promoted by the Amazon Cooperation Treaty and international cooperation) of defining criteria and indicators for sustainable forest management. In many countries, national and state ministries, in addition to departments and even municipalities, prepared plans for the use of public lands and forests to promote sustainable development combining forest use with conservation. Generally, the plans arose, at least partially, from discussion processes and the active participation of civil society in public forums and events (EFTRN, 2004), often based on information gathered through extensive natural resource and socio-economic inventories. In addition to providing descriptive facts and statistics for the national forests, these programmes also provided a focus for determining the main course of action for the land and natural resources, considering issues such as the type of use, priority actors and protected areas.

In Brazil, for example, the inter-ministerial National Forest Programme, implemented by decree in 2000, articulated and directed the actions of the Brazilian government in relation to forest resources. This programme was coordinated by the Ministry of the Environment and executed with the support of CONAFLOR, a governmental organization specifically created for this purpose. As in other countries, this process was co-funded by international donors, which in this case included organizations such as the International Tropical Timber Organization

(ITTO), the Pilot Programme for the Protection of Tropical Forests (PPG-7) and the Global Environmental Facility (GEF).

In several instances, in the operationalization of the directive strategies defined in the national forest programmes, whether at the state or municipal level, many governments also undertook an Economic-Ecological Zoning process for their Amazonian territories. Based on information about the biophysical environment and the socio-economic situation, the zoning process determined spatial directives regarding the use of resources, or restrictions thereof, aimed at sustainable development. Often, this process included profound background studies of selected ecosystems or on the relationships between society and the environment. Thus, zoning was one of the main environmental policy tools with spatial relevance that supported policy-makers in their decisions about which land uses to promote, which areas to exploit and (especially) which to protect and where to concentrate efforts for law enforcement. For example, the Economic-Ecological Zoning process defined the economic activities (agricultural, forestry, industry, etc) that may be performed or should be developed in certain areas. Additionally, the zoning process was an important tool for negotiations between the government, the private sector and civil society regarding possible alternatives for sustainable regional development. In particular, this process provided the opportunity to include representatives of the locally relevant landowners who held the rights to exploit resources and, therefore, had the motivation and power to develop the economic, ecological and social potential of the region. In Brazil, one of the most advanced countries in this initiative, the Economic-Ecological Zoning process for the Amazonian States was concluded in 2011.

The elaboration of the plans and supporting studies had important societal effects (see, for example the FAO, National Forest Programme Facility: http://www.nfp-facility.org/en). The process enhanced the visibility of the environmental sector in the public policy arena and generated a platform for exchange and discussion between governments, the private sector and civil society on regional development options. It therefore increased the knowledge and recognition of the decision makers regarding the importance of the environmental sector and the associated social and economic implications. However, the observations in the study regions revealed that these strategic planning efforts had mixed consequences for the local population. Thus, although the importance of the forests for local people was often recognized, in the majority of the final decisions, environmental and economic concerns were consistently given more emphasis than social considerations, as the multi-stakeholder dialogue was strongly dominated by environmental NGOs and by representatives of the private sector (Pacheco et al., 2008). In the Brazilian state of Acre, for example, 50% of the territory was designated as protected areas as a result of the planning process. An additional 400,000 hectares of degraded areas and another 400,000 ha identified as being in the process of degradation were deemed unsuitable for economic activities, so the smallholders' options for using natural resources were strongly affected in the long run. Land-use planning programmes could easily escalate conflicts over land and resources, negatively affect options for using natural resources in those areas classified for conservation and facilitate the timber companies' access to forests traditionally used by local families.

Perhaps the most relevant result of these strategic planning efforts for part of the local population was the gazetting of large protected areas designated as public property with restrictions on the utilization of natural resources, particularly the forests. This type of area can be divided into areas protected for environmental purposes, areas of exclusive use for the local population, indigenous reserves dedicated to the protection of ethnic groups and public areas with a focus on geopolitical or military ends. Similarly to Nelson and Chomitz (2009), we found that, in the ForLive study areas, the protected areas were an effective tool for protection against deforestation, especially when combined with an intensification of public control (see also Plate 12). However, there was evidence that actors may compensate for the protection of one area by increasing the intensification of use in other areas that are less protected (Porter-Bolland et al., 2012).

The emphasis on the environmental dimension in these strategic planning efforts, in addition to seriously challenging the very limited capacities of the governmental authorities for enforcement, had mixed effects for the families that depend on the resources regulated by this protection status. For example, in the 'Verde para Sempre' Extractivist Reserve, located in the municipality of Porto do Moz (Brazil), immediately after the creation of the reserve in 2004, illegal timber harvesting massively decreased (Martins et al., 2007). The new status as a protected area, combined with the greater presence of governmental authorities and diverse NGOs, did indeed protect the area to a certain extent against the invasion of and illegal use by external actors. Consequently, the establishment of the reserve not only improved the protection of the environment, but also reinforced the position of the families and their traditional rights to resources. However, there were also negative effects for the families. Suddenly, they were forced to enter into bureaucratic processes to legalize their traditional ways of using their natural resources. The use of their own forests, for example, required authorized management plans, whereas other land uses, such as the herding of buffalo, became illegal. Another negative impact on the smallholders was that the legal framework for Extractive Reserves generally prohibited the sale of individual properties. Even in 2012, eight years after the creation of the reserve, the responsible authorities remained unable to elaborate a formally recognized strategic plan for the management of the reserve, although this plan was a prerequisite for the legal use of resources by the local families.

The bureaucracy involved in authorizing the traditional use of resources caused many difficulties and much resentment against the protected reserve status. Although it supports the conservation of resources and the protection of families in unfavourable negotiations with external actors interested in their lands and resources, this system also reduced the flexibility and freedom enjoyed by these families and often made their daily life more complicated. In this context, it is not surprising that, despite the new norms and restrictions, the families in this study area continued to engage in the illegal sale of timber to companies and the sale of land, even without the proper legal qualifications to do so. These observations indicate that the authorities, in collaboration with diverse NGOs active in these new contexts of legally protected areas, were not capable of implementing consolidation and coordination of activities in the legal reserve, nor did they develop viable economic livelihood options for the smallholders. This phenomenon appears to be typical of a large majority of the protected areas recently created throughout the study regions. The delimitation of protected areas, although it is an important step in the right direction, generally remained an isolated action with results that, until now, have had mixed consequences for the families in and outside protected areas (Ehringhaus, 2005).

Improving the legal and institutional forest framework

Over the last few decades, following Rio Summit in 1992, almost all the Amazonian countries have made significant efforts to improve legal-institutional frameworks for forest use. This process generally received massive support through international collaborations that sought to consolidate the regulatory and institutional framework as a crucial step in ensuring sustainable use and conservation of forests in the region. The best-known examples of these efforts are likely the legislative reforms regarding forests in Bolivia, which were developed within the framework of the BOLFOR project, primarily financed by USAID, and the forestry legislation reform in Brazil, which was undertaken with the support of the Pilot Programme for the Conservation of the Amazon (PPG-7), largely dependent on support from a German cooperation project.

The evolution of the normative framework

A comparative study of the legal and institutional framework relevant to forest use by smallholders in the four countries included in the ForLive project (Sabogal et al., 2008a; Carvalheiro et al., 2008; Ibarra et al., 2008; Martinez Montaño, 2008) revealed that all countries generally followed the same reform model, which is perhaps most typically reflected by the evolution of the legal-institutional framework in Bolivia. Figure 5.2 schematically illustrates the design of this general model of legal-institutional reform over time, indicating the effect of the various steps of reform on forest use.

In many countries, there was no legal basis for the use of forests in the Amazon until the last few decades. As a result, there was mostly only informal use by local actors, often quite extensive and based on simple traditional rules (Pokorny et al., 2009, 2010b). Occasionally, companies utilized loopholes to exploit high-value tree species such as mahogany or for the large-scale transformation of forests for other land uses, mainly cattle (Bunker, 1985). With an increasing perception of the strategic environmental and economic importance of the Amazon region and its forests – and the related integration of the region into the national economy – governments started to create environmental laws and, within these laws, specific regulations related to forests that defined more strategic objectives for their use and protection. In general, these laws suffered and still suffer from significant

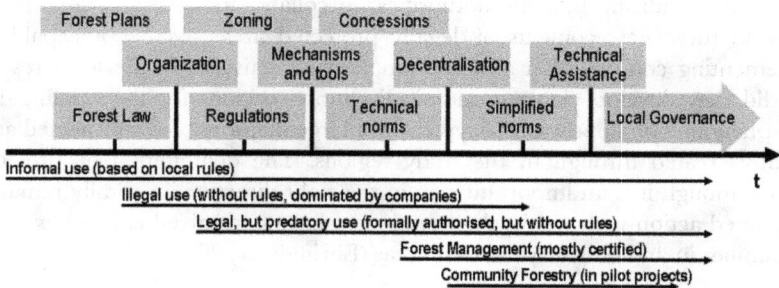

Figure 5.2   The evolution of the legal and institutional framework for 'Community Forestry' in Amazonian countries and its effect on local peoples' forest use

incompatibilities with the regulations of other sectors, especially agriculture, mining, transport and (hydroelectric) energy (see Chapter 7). The underlying conflicts of interest between these sectors still cause severe inconsistencies in regulations and policies.

Thus, Amazonian countries began to specify, sometimes over very long periods, when and in which way land and forest users were allowed to access the forests. In this phase of regulation, due to the lack of sufficiently specific rules and the general absence of effective mechanisms for law enforcement, companies and smallholders continued to use the forests as they had been doing previously, without regard for the changes in legislation, with the exception of several formal requirements for authorization such as presenting the results of an inventory or a management plan signed by a forest engineer. Control of the information and contents presented in these plans, or even audits in the forest management areas, seldom occurred. Many countries in the region remained 'stuck' in this ineffective regulatory state until the 1990s, when they finally began to create more specific technical norms and establish an organizational framework to enforce the regulations more effectively.

The first generation of these technical norms originating from the decisions of the Rio Summit was strongly oriented towards the exploitation of timber by companies, partially because it was based in the practices of Reduced Impact Logging (RIL). Correspondingly, joint international and national efforts were invested to strengthen environmental competency and skills at all governmental levels, from the ministry to the prefectures, particularly through three actions: (1) the creation of competent agencies responsible for planning and controlling the use of forests, generally affiliated with the environmental departments; (2) the improvement of technologies and procedures for control; and finally, (3) attempts to decentralize administration through the transfer of administrative responsibilities from a central government to governments of the state, local departments and municipalities.

Another crucial element in this phase of operationalization was the demarcation of forest concessions, that is, public forest areas made available for harvesting by timber companies in exchange for the payment of royalties. Concessions are normally distributed in public auctions. Often, the demarcation of these forestry concessions was linked to efforts for zoning and land-use planning. As such, the attempt to more effectively control the use of forests occasionally contributed indirectly to the recognition of local land and user rights. In many cases, however, we found that these efforts simply demonstrated and reinforced the existing conflicts for land and resources, and in the extreme, they allowed the timber companies to harvest in forests traditionally used by local families whose informal rights were simply ignored. Thus, on the one hand, the principal impact of the stronger regulation of the forestry sector was the emergence of a form of legalization of predatory timber harvesting operations characterized by the existence of legally approved Forestry Management Plans and the intentional disregard for these plans by the companies, which simply continued with their predatory harvesting operations without being prosecuted or even controlled (Sabogal et al., 2007). On the other hand, a few companies were partially mobilized by enhanced legal regulations and gradually improved institutional control. These companies invested in the FSC certification of their forestry operations based on the practices of Reduced Impact Logging. Nevertheless, few companies were able to comply with the RIL standards (Pokorny and Steinbrenner, 2005; FSC-AC, 2010).

As a consequence of these shortcomings, the difficulties and ambivalent implications of the forestry policy framework with regard to mechanized commercial timber management by professional companies became increasingly apparent. Thus, more recently, governments have made an effort to create space for more social participation by local forest users, managers and owners, particularly including organizations representing indigenous groups, traditional communities and settlers. This process aimed to give the aforementioned groups more influence in decision making and to make the regulatory frameworks more suitable for smallholders, a trend reflected in various global discourses (Arts and Buizer, 2009) including sustainable development, self-regulation and global environmental governance, and in the increasing global interest in fighting poverty evident in the organization of the World Summit in Johannesburg and the declaration of the Millennium Development Goals (MDGs). In the new forest legislation that was enacted in many tropical countries during the 1990s (Silva et al., 2002), non-traditional forest users who had gained territorial control over forest lands were now accommodated. Moreover, there were attempts to also consider NTFPs in the regulations.

Again, a host of international donors and national NGOs entered the region, this time, focusing on the promotion of community forest management (Pokorny and Johnson, 2008[C], Pokorny et al., 2012). The insights and experiences gained within these initiatives and networks gave rise to a series of negotiations to change and adjust the legal-institutional framework to make it more suitable for smallholders, whose realities differed significantly from those of the initially targeted companies. Certain countries were more advanced in simplifying the regulations to reduce the bureaucratic procedures for smallholders, which encouraged the smallholders to legally use their forest resources for commercial purposes. In

Ecuador, for example, an instrument for simplified management plans (Planes de Aprovechamiento Forestal Simplificados; PAFSi) was established, and in Peru, the relevant governmental department (Instituto Nacional de Recursos Naturales; INRENA) distinguished three levels of intensity in the authorization of forest use for indigenous communities depending on the area and volume of the forest being used and the socio-economic features of the community. There were also several attempts to establish technical assistance programmes for 'Community Forestry', for example, in Ecuador and in the state of Acre, Brazil.

However, we found, as did Larson et al. (2006) and Pacheco et al. (2011), that the advances in influencing local public politics in the study areas were only moderate and did not necessarily translate into improved access to resources and other financial and physical assets by smallholders. The limited capacities of the public administration, the insufficient law enforcement capacity of the state and lengthy and extremely bureaucratic processes significantly limited the ability of smallholders to use the regulations to their benefit (see Box 5.1).

During the study period, Amazonian countries were all in different stages of this legal-institutional process. Several governments continued seeking opportunities to better adapt to and consider smallholders in their forest regulations, whereas others were strongly focused on the demarcation of concession areas assigned to timber companies with sufficient capital and knowledge to comply with the legal requirements. Although the majority of the countries invested in improving and professionalizing the administrative and technical support of local management plans, there were also several cautious attempts to explore possible alternatives for a more effective consideration of social issues by giving more autonomy in the design of the management scheme to the communities themselves.

Box 5.1 *Efforts and problems in using decentralization to facilitate the legal use of forests by smallholders in Ecuador*

Since 1999, through the Federal Law of Decentralization and Participation (Government of Ecuador, 1997), the Ministry of the Environment has initiated a process for the transfer of administrative functions and supervision of headquarter regions to civil society. In this aim, a National System of Outsourced Forest Monitoring (Sistema Nacional Tercerizado de Control Forestal) was created; this system integrated three levels of public and private operations to complement the work of the forest authority: green supervision ('Vigilancia Verde'), conducted by a semi-private forest control group that operated through control points on timber transportation routes; forest stewardship ('Regencia Forestal'), which delegated the work of supervising authorized forest management operations to forest engineers accredited by the Ministry of the Environment; and a private company (SGS) that administrated and supervised the verification system for forest operations (Arias et al., 2006).

In 2006, a consultancy process was initiated to promote a control system that culminated in the establishment of the National System of Decentralized Forest Monitoring ('Sistema Nacional Descentralizado de Control Forestal'; SNDCF). The Ministry assumed a governing-body role that defined the forest policies and coordinated the SNDCF at a national level. In this system, the provincial governments took responsibility for the authorization of forest management plans, transport permission for the harvested timber and verification of the correct execution of harvest operations in the field. The National Ministry of the Environment controlled forest administration in the diverse regional districts, technical offices and decentralized jurisdictions (MAE, 2006[A]).

Unfortunately, the SNDCF was never implemented because none of the provinces were able to comply with the necessary requirements to establish such a system.

Obstacles for smallholders in the use of their forests

Several studies (Sabogal et al., 2008a; Carvalheiro et al., 2008; Ibarra et al., 2008; Martinez Montaño, 2008) examining the perceptions and experiences of smallholders with regard to forestry policies showed that, in all the study areas, the simplification of norms and greater access to technical assistance were not sufficient to overcome the principal obstacles faced by smallholders and communities in achieving the legal use of forest resources. Based on workshops attended by representatives of governmental organizations, NGOs and local organizations, the study identified the persistence of a large number of obstacles that impeded attempts to achieve a better legal and institutional framework that could effectively facilitate smallholders' opportunities for the legal use of their forest resources (see Table 5.1). The discussions revealed that the issues related to the process of land regularization and tenure represented the greatest obstacle for smallholders' effective use of the forests. It was widely accepted among various stakeholders that ensuring ownership and subsequent access to forests is a key step in promoting real progress in the life of the smallholders. It also became obvious that the vast majority of smallholders simply did not know the laws and regulations. In combination with a huge lack of technical assistance and the insufficient law enforcement capacity of the state, this situation significantly limited the ability of smallholders to use the regulations to their benefit. There were also significant inconsistencies in the regulations and definitions, particularly with respect to the legal frameworks of other sectors. Although the process of institutional reform contributed to better implementation for governmental organizations, the application of the regulatory framework was generally quite poor, with very limited efficiency. The workshop participants particularly stressed the lack of institutional competence and extremely bureaucratic and lengthy processes, entangling smallholders in their attempts to use their resources legally.

Table 5.1   Problems relating to the legal-institutional framework for forest management
by smallholders in the Amazon

Context
Land tenure issues not resolved
Large influence of the private/business sector
Lack of information on laws and regulations, markets, support opportunities (e.g. credit
and technical assistance), etc
Legal framework (Norms – Regulations)
Homogenization and simplification of social actors
Strong emphasis on mechanized utilization of timber
Many technical demands and requirements and bureaucratic administrative processes
Cultural incompatibility
Incoherence with other sectors
Judicial deficiencies (e.g. inconsistencies or legal 'vacuums')
Institutional framework (Organisations – Actors)
Weak operational capacity of the government institutions
Limited visibility and understanding of the defined responsibilities of the diverse
governmental organizations
Centralized administration
Significant bureaucracy and complexity involved with the procedures, lack of
transparency and delay in the conclusion of processes
Controlling attitude
Lack of support mechanisms
Focus on the authorization and auditing of forest use rather than on the supervision of
activities and control of illegal use
High frequency of changes in the procedures and rules

The standard regulatory frameworks employed a simplistic and standardized vision of forest users' cultural, biophysical and socio-economic reality. Consequently, they generally suffered from systematic problems: the continuing incompatibility of the simplified norms with local realities and the persistent requirement to enter into bureaucratic processes. The new regulations also widely failed to adequately consider the large diversity of smallholders' forestry and agroforestry practices, which include not only timber but, often much more importantly, NTFPs and environmental services. In Peru and Bolivia, for example, the simplified regulations even banned the use of chainsaws for the pre-processing of logs in the forest, which is realistically the only way that small forest users can drag the timber out of the woods on their own without depending on the heavy machinery of commercial loggers, and thus ensure a fair share of profits from the commercialization of their timber resources.

Furthermore, with respect to governmental efforts to control the actions of the private sector, we found evidence of serious shortcomings. Governments were rarely successful in establishing viable mechanisms for controlling and supervising the powerful actor groups, even when there were clearly defined regulations, as in the forest sector (Sabogal et al., 2008b); however, the states were able to identify environmental crimes in remote rural areas of the Amazon thanks to the application of advanced remote sensing technologies to monitor illegal forest use, which were recently implemented by many countries with substantial international support. The states still lacked the tools and resources to apprehend and effectively punish those responsible for illegal use. Like Brito and Barreto (2006) and IMAZON (2005), we found many situations in which the authorities had identified the companies or people responsible for environmental crimes but had no way to punish them; some officials simply neglected to prosecute such crimes due to corruption or tight (often personal) political links with the delinquents (Parker et al., 2004; Thomas, 2008). This lack of effective control allowed the timber industry, ranchers and agribusiness to easily advance the illegal use of land and forest resources. Based on these occurrences, it can be concluded that authorities must be more effective in dealing with the private sector in order to redress inequalities and protect smallholders from illegal exploitation and unfair settlements.

Local governance

Considering that the large majority of smallholders have a limited ability to meet the demands of the extant legal-institutional model (even in a simplified form) without significant support from external organizations, and given the structural difficulties that authorities encounter in enforcing legislation, alternatives are being sought. In this context, the possibility of delegating power and responsibility for the control of natural resources to local institutions utilizing the capacities at their level has been emphasized (Ostrom, 1999). In particular, a large number of successful experiences with informal local agreements on the use and management of common resources have been noted as a possible starting point to overcome the difficulties associated with the implementation of externally defined and enforced models and technology transfer. In these local agreements, the smallholders themselves managed to negotiate and agree on the rules for access to and the use of resources of common interest to ensure the maintenance and quality of the resources and their effective use over time. Various positive experiences have demonstrated that, when local communities govern access and use of resources in accordance with their own rules, they are more likely to implement and ensure compliance with these local rules (Guijt, 2007). These local governance schemes also allow for a better, more flexible and more immediate adaptation to the dynamic context typical of the region, as they can be discussed and adjusted directly by the natural resource managers without depending on externally governed decision-making processes. Thus, these systems tend to strengthen the local organization and social capital of the communities, reinforcing the role of local populations in society and strengthening the local societal base (Vázquez-Barquero, 2002).

In all the study areas, we found examples in which the communities established procedures for the regulation of access to and use of local resources, including the demarcation of forest areas for common use, the establishment of norms for the utilization of rubber and Brazil nuts, and fishery agreements (see Box 5.2). The analysis of these examples of local governance schemes, which often also included regulations on the use of the resources by external agents, showed that they typically emerged when external actors such as ranchers, loggers or commercial fishermen, without previous negotiation, started to systematically exploit the resources of the local communities. In fact, in all the successful examples of local governance observed in the study areas, it was when such threats to their livelihood arose that the families organized themselves around their representative organizations, including syndicates, committees and associations, to reinforce their interests and rights against outside threats (Medina, 2008).

In the case studies, the demands of the local community and their representative organizations received scant attention in the political arena outside of the communities. Although highly functional and valid, the majority of these local rules were not supported and recognized by the governmental authorities. The communities' systems of management were only recognized when local families managed to establish alliances with powerful external actors, particularly with the environmental NGOs. However, in exchange for their support, these allies directly or indirectly pressured their local partners to adapt to their interests and discourses (Medina et al., 2009a). Thus, it was not the establishment of local rules and mechanisms for planning, managing and controlling resources, but rather the institutionalization and legal recognition by relevant authorities that constituted one of the main challenges of this alternative model.

In general, the findings of ForLive confirmed the observation of Pacheco et al. (2008) that, despite the proliferating discourse on the importance of more participation by local actors in public policies and the governance of forests and other natural resources, there were very few governmental initiatives that were able to resolve the conflicts of interest between local people and external actors. Moreover, governments rarely found ways to successfully address the related legal challenges in handing over responsibility, control and resources to local representative organizations. The immense problems facing the Bolivian government in its attempt to establish a system of legal pluralism that strives to recognize not only the usual state jurisdiction, but also the traditional community systems is an example of the magnitude of these challenges (Romero, 2011). Although this approach was previously explicitly incorporated into the Constitution, it remains largely unclear how such a system might function in practice. Another example of the difficulties of embedding local governance systems in the formal regulatory framework is that of the management board of conservation units in Brazil (Sistema Nacional de Unidades de Conservação da Natureza; SNUC), conceived as panels in which representatives of all relevant actor groups can make decisions regarding the unit in question. As these boards must also act according to the applicable laws and regulations, the decision-making process is often paralyzed, and the decisions that are made are often difficult to implement, as they are blocked by the government due to legal concerns, differing political positions or simply incompetence.

Box 5.2 *Local fishery agreements in Porto de Moz*

In the 1980s, in the Xingu river network within the municipality of Porto de Moz, fishing companies from the cities, employees of small timber companies and even local fishermen began to systematically use fishing equipment such as fishing nets, lanterns and diving masks to increase their effectiveness. As a result, many of the rivers, generally those parts most attractive for fishing, became overfished. This caused a drastic reduction in fish populations, particularly among the species crucial for the basic livelihoods of the local river communities. Because local families suffered disproportionately from this situation, several of them started initiatives for the regulation of fishing in their rivers. They discussed norms to regulate access to fish by commercial fishing companies organized in the local fishermen's association and by families living in the communities. The norms defined parameters such as periods for fishing, the minimum size of the fish and equipment regulations. The discussion of norms initially caused some internal conflicts among the communities. However, these communities gradually began to achieve consensus. Some agreements were not written down but were transparently discussed in meetings and etched in people's minds. Several other communities wrote down and formally signed their agreements. Thus, among these families, breaking the rules meant violating the rules of coexistence.

Communities with fishing agreements	Rivers
Cupari, Maria de Mattias, Vila Nova Bom Jesus, Bom Jesus; São João do Cupari	Coati and Cupari
Espiritu Santo	Açaí
Arurubarra	Aruru
Monte Sinai, Santa Luzia Cuieiras, Miritizal	Uiui and Peituru
Cajuí	Amazonas
São José, Espírito Santo, Santa Luzia, Seguidores de Cristo	Majari
Acorda para o Lago do Urubu	Uiui

Although Brazilian legislation allowed the relevant environmental agency (IBAMA) to recognize this type of agreement and even specified modalities to do so, the agreements developed by the communities in Porto de Moz were not legally recognized for a long time. The justification was that they did not meet the minimum technical and scientific criteria. The lack of normalization and acceptance of these agreements weakened and demobilized the communities. However, even given this dynamic, the agreements helped the communities maintain a certain amount of control on fishing activities.

▶

Finally, following the ForLive project, the local representative organization (the Committee for Sustainable Development; CDS) successfully negotiated the formalization of several of the local agreements with the Fishing Secretary of the State of Pará

As described above, the discussion on the operationalization of the concept of local governance in the Amazon region is still in its initial phase. Despite the widely acknowledged potential, there are many important obstacles and difficulties related to the implied necessity of adjusting the legal framework in a way that permits the transfer of state powers to local organizations of a private, local nature. Additionally, there is a major difficulty in justifying a different treatment of smallholders in comparison with other actors, who cite the constitutional right of equality under the law, as differentiating smallholders implies a degree of favouritism. In addition, there were again serious concerns that powerful actors with access to capital could find ways to take advantage of emerging situations for their own benefit.

6 Experiences with local development initiatives

The many efforts made to support smallholders naturally affected the life of families in the region. This chapter presents the specific positive and negative effects and the benefits and difficulties related to these initiatives for development. In the case studies, we found evidence of improved well-being for many families, but there were also several negative effects with implications for the region's future. In particular, the study revealed that development initiatives, as currently implemented by public and private organizations, generally accelerated the process of homogenization of the huge diversity of cultures and locally developed productive systems found in the region.

Benefits of the external development efforts to support smallholders

During the project, the numerous discussions held with all actor groups in the study areas made it clear that the vast majority of smallholder families in the region would not be able to reach an acceptable level of well-being on their own by relying solely on the management of their natural resources. In fact, many of the families in the region lived below the poverty level and suffered from major problems in terms of nutrition, health, access to energy and education (UNDP, 2010). The indices of human well-being in the study regions were generally low, typically lower than the corresponding national average. The precarious situation in which many families lived was identified as one of the principal reasons for the massive out-migration of people to the rapidly growing urban centres or, particularly in the Ecuadorian study area, abroad. In a certain sense, many families living in the rural Amazon were virtually excluded from public services and consumption markets. The weak presence of the state in the more remote areas, combined with a high level of corruption within the few established agencies, left the smallholders highly vulnerable to the actions of powerful actors who effectively satisfied their own interests (Parker et al., 2004; Thomas, 2008; Weigelt, 2011). In all the study areas, we observed conflicts, often violent, between settled families and powerful newcomers. Often, the families were not able to exercise their rights, even when such rights were legally acknowledged. In fact, the vast majority of the families were not even aware of the existing regulations or their legal rights (see also Chapter 5). In general, the smallholders in the study region participated only marginally in public policies, as exemplified by the fact that the local families' representative organizations – such as worker associations and social movements – were mostly opposed to the ruling district governments, in which they only occasionally participated as junior partners. This situation was further aggravated by the fact that a large proportion of the families, particularly those living in more remote rural areas with extremely low incomes, generally did not feel represented by their organizations, which also often acted on the basis of political and relational representation.

In these contexts, we found hundreds of development initiatives implemented at the regional and local levels with the help of governmental organizations, NGOs and socially involved individuals. The great needs of the rural families and a general belief in the opportunities presented by these organizations, which were perceived as rich and powerful, created a dynamic wherein the families were enthusiastic about whatever proposal or project the development organizations proposed. In fact, in all the case study areas considered by ForLive, no examples were observed in which a family or community had entirely rejected a proposal made by an external organization, regardless of the specific nature and objectives of the initiative. As a result of these projects and of policies that considered and protected the human rights and necessities of the families and, most importantly, the efforts of the smallholders themselves to have their rights acknowledged (Medina, 2008), conditions for many of the families in the study region improved significantly. When comparing their current situation with conditions during the post-colonial and feudal times – still perpetuated in more remote regions – the families reported a significantly higher level of well-being including regular meals, access to public health services and protection of their human rights (see Box 6.1).

Box 6.1 *The outcomes of development (reflection by Vincent Vos, scientific assistant in ForLive)*

Although there is evidence that, in pre-Columbian times, great civilizations lived in the Bolivian Amazon, the diseases brought by Europeans combined with population movements caused by the aggressive invasion resulted in a severe reduction of the native population. From the very large original diversity of indigenous groups existing in the Bolivian Amazon approximately 100 years ago, there are only a few semi-nomadic indigenous groups left today (Denevan, 1992; Mann, 2005; Balee and Erickson, 2006). Northern Bolivia was one of the last areas of contact. With the growth of the rubber market, large business and adventurers arrived from La Paz, Santa Cruz and other regions and appropriated large tracts of land with little respect for the indigenous and local populations. Generally, these newcomers contracted people from other regions to work as rubber tappers in a semi-slavery system based on false promises and indebtedness. In many cases, these businessmen and their workers killed many indigenous inhabitants of the region, taking their women. Today, the descendants of these relationships, the mestizos, make up the bulk of the regional population.

Often, the attempts of the indigenous people to resist this abasement were answered with cruel counterattacks, even resulting in the genocide of several of these groups. For example, after the 'Pachauaras' killed rubber tappers in the Orthon river region, the businessman Antonio Vaca Diez commanded a brigade to attack them, sparking a clash that is the principal reason why there are only eight 'Pachauras' left today. Several other ethnic groups such as the 'Araonas', the 'Chacobos' and the 'Esse Eja' managed to escape these cruelties by going deeper into the forest. Others, such as the 'Cavineños' and 'Tacana', integrated themselves into the feudal extraction systems, often catalyzed by the work of missionaries who, over time, adopted the bearing of the rubber barons, including renting and selling indigenous people as property.

Although the fall of rubber prices in the early 1920s naturally abolished much of this system of patronage and semi-slavery, the main components of this system still persisted in the region until recently. For example, indigenous leaders told me that, until the last decade, Fredi Hecker, a rubber baron with great political power, candidly announced that there could not be any indigenous people living in the region who might have rights over land and resources because his grandfather had killed them all. The Heckers owned approximately 300,000 ha of land, geared predominantly towards the production of Brazil nuts, where thousands of descendants of these families were working in a form of semi-slavery, meaning that they received little payment while being forced to pay for expensive housing and food.

In fact, only in 1997 did the government officially recognize the existence of indigenous villages in the northern Bolivian Amazon in response to a series of protest marches in La Paz organized by various social movements. In the wake of more protests, hunger strikes and five more marches, these lands were finally withdrawn from the barons with the implementation of legal land reforms. The illegally appropriated land was then redistributed to the families who formerly worked there, mainly indigenous people and peasants. Consequently, many of the large land owners lost their land (the Hecker family, for example, retained only 1,562 hectares according to the information provided by the Ministry of Land), and large tracts of land were assigned to peasants and indigenous people, who finally had a guarantee that no one would be able to take the land away from them or demand part of their production. Smallholder communities, especially those located close to urban zones, were able to organize themselves with the support of various NGOs, significantly strengthening their negotiating power. For example,

▶

the price received by these communities for a box of Brazil nuts (23 kg) increased from Bs10 to Bs120 over a period of 10 years (Rodriguez, 2008). Several communities were also able to negotiate with the local authorities for the construction of new roads, schools, offices, sports fields, sheds and houses. With the support of development organizations, these communities are increasingly improving their production systems and their quality of life.

Although poverty continues to be very visible, important indicators such as the rates of malnutrition and infant mortality decreased significantly, whereas the producers' self-esteem and respect for their rights has increased. However, there remain many opposing political and economic powers that impede local development (the massacre of peasants in Porvenir shows the lengths to which these powers are willing to go to maintain their hegemony). However, the new national government of Evo Morales has begun to establish a better presence in the region, which, in combination with the changes in property rights and the empowerment of the local populations, indicates that producers, peasants and indigenous people have succeeded in escaping the tyrannies that they faced for centuries and have initiated a process of self-development. Keeping in mind these enormous advances, I do not think that it would be fair to discuss only the problems resulting from the development initiatives.

In all the external initiatives for local support that were studied in ForLive, the families demonstrated the capacity to comply with the requirements of the technology and management packages proposed by the development organization. The smallholders were able to apply the suggested techniques and establish the proposed forms of organization. However, it became obvious that this implementation was only feasible with significant external support including training, payments and technical advice. Generally, these programmes were oriented towards a more continuous accompaniment, and the organizations working with well-trained local staff were more successful in supporting the families (Hoch, 2009). In almost all the initiatives, at least some families were able to take advantage of the opportunities offered by the programmes and development projects (Hoch, 2009) to improve their situation. Often, the collaboration helped to make the commercialization of local products and the use of available natural resources more effective and profitable, particularly in projects dealing with forest management and agroforestry (see Box 6.2). As previously noted by Humphries and Kainer (2006), our studies showed that families participating in development initiatives acted more consciously, communicated more effectively and were successful in leaving their isolated lifestyles, as they began to collaborate more intensively with external actors and participate more actively in the formulation and processes of public policies.

Box 6.2 *The important role of NGOs in disseminating agroforestry systems in northern Bolivia (adapted from Escalera and Oporto, in preparation)*

For more than a decade, the small Bolivian NGO Instituto para el Hombre, Agricultura y Ecologia (IPHAE) has been active within an area of approximately 750 hectares of agroforestry systems in 45 rural communities in the region of Riberalta, providing benefits to more than 700 families. A study comparing three families with support and two families without support in the community of Palmira showed that the families who followed the advice of the NGO significantly improved their income and rose above the poverty line. Moreover, the agroforestry systems implemented here facilitated the partial reclamation of highly degraded land and thereby contributed to environmental conservation. In view of this success, some families also tried to adopt the new technology without support from the NGO but failed. Commenting on these failures, Mr Saul, one of the successful producers participating in the project, said 'planting and maintaining some trees for subsistence is easy, but if one wants to commercialize them, he or she needs technical support'

Difficulties

The many development initiatives in the region that were directed at small-holders created an opportunity to break from the historical relationships based on exploitation by external actors and to alleviate some of the socio-economic insecurity in these communities (Sabogal et al., 2008b). However, it is important to keep in mind that these efforts to enable smallholders to participate more effectively in the opportunities emerging from development dynamics necessarily impact the social system. As with all interventions, in addition to the generation of desired, positive effects, there are also unintended consequences, both positive and negative. The development initiatives observed within ForLive confirmed the potential to strengthen the local social system and the local capacity to adequately respond to opportunities and threats. However, we also observed the potential for adverse effects that were often overlooked by the supporting organizations. Naturally, these organizations are interested in claiming success and therefore tend to analyze and describe the positive effects of their interventions, whereas possible negative consequences may not be adequately perceived or dealt with, a phenomenon described by Rogers (2003) as 'innovation bias'.

With the goal of evaluating possibilities to adapt and optimize external efforts for local development, in this section, we reflect critically on the difficulties and possible negative consequences of development initiatives by exploring the insights gained in the case studies. First, we address the difficulties for the smallholders who participated actively in the development projects; further, we examine the broader effects of development initiatives that target smallholders in the region. In general terms, this analysis reveals that the development approach, as currently promoted by the vast majority of development organizations, improved the living conditions of some of the families but in general terms also contributed to further marginalizing smallholders' original socio-productive systems. Thereby, development projects indirectly advanced the general development dynamic characterized by cultural homogenization, the functional accumulation of land and resources and, in turn, environmental degradation.

Inside the pilot projects

Almost all the ForLive case studies that were dependent on external support faced severe difficulties. Almost none of the families participating in the studied initiatives in 'Community Forestry', agroforestry and plantations managed to adopt the proposed technologies and procedures without continuous support (Pokorny and Johnson, 2008b; Medina et al., 2009c). This observation directly contradicts the ideas inherent in many development projects suggesting that, after initial support is given to the smallholders, they can then continue production on their own after a period of consolidation. Moreover, we found that the projects had many ambivalent effects on the social systems of the families and communities.

Withdrawal and lack of spontaneous replication

All the local economies that we found in the rural zones of the study areas were traditionally based on diversified production systems aimed at satisfying local consumption needs and using occasionally emerging opportunities for the commercialization of specific agricultural and forestry products, often organized by intermediaries and powerful families, and oriented towards nearby urban centres. These economies were mostly self-managed by the smallholders and other local actors and, as a direct consequence, were in line with the existing local capacities, institutions and contexts. The proposals of the majority of the development organizations analyzed in the project inserted a very different economic model into these local contexts (Porro et al., 2008; Pokorny et al., 2012). Although the majority of the interviewed supporting organizations explicitly highlighted the importance of considering the specific culture and local knowledge of the local counterparts, their proposals were almost always based on legal and scientific/ technical criteria formed with little input from the smallholders; moreover, the proposals often even ignored locally existing capacities. Thus, local management practices, the local way of organizing work and the capacities and limitations existing in the localities were largely ignored in favour of a new and 'improved' way of working originating from a completely different knowledge set (Porro et al., 2008; Pokorny and Johnson, 2008b). These proposals for development were implemented under a business-oriented vision that usually required new organizational skills, included the application of more sophisticated technologies and depended on higher resource input (see Chapter 5). Consequently, for all the families contacted by ForLive, meeting the demands of the technology packages and organizational suggestions made by the development organizations represented an enormous challenge, and they faced severe difficulties in adapting to the demands of time, technologies and organizational schemes that were being promoted.

The costs of implementation and maintenance of the new technologies and organizational schemes were high and typically exceeded the existing local capacities and resources. In fact, many of the analyzed development projects required human resources and financial support that were completely incompatible with the realities of the families (see also Figure 5.1). In the studied cases, we observed a tendency of smallholders to adopt the external proposals only in the initial phase of the project, when the supporting organization guaranteed the availability of the necessary funding, know-how and equipment. However, particularly in the case studies where there was high initial donor support, the families became extremely critical after the intensity of support decreased and they were expected to cover the operational costs on their own. In nearly all the case studies, this phenomenon of increasing scepticism was further aggravated by the fact that the outcomes fell far short of the benefits originally promised by the external organizations to mobilize the families for cooperation, whereas the inputs and costs were higher than expected. Another critical factor was that, in their attempt to establish connections with financially attractive national or international markets, the development organizations induced more complex, often non-transparent, marketing schemes, which provoked significant delays in payments for the products, even though immediate returns were generally of great importance for the families (Hoch, 2009). Consequently, the majority of families decided to resume their traditional management schemes after the termination of external support, so the level of long-term adoption was low. We found that many families who had participated in a development initiative became more sceptical about development projects.

The problem of the lack of long-term adoption of innovative technologies by smallholders is illustrated dramatically in Figure 6.1, which shows the rates of adoption observed for tree plantation initiatives, which were typically implemented over the course of decades by public or privately financed programmes (Hoch et al., 2008; Hoch, 2009). As described in Chapter 5, in the majority of these programmes, the supporting organization simply distributed plants to the families and provided some technical support and, less frequently, financial support during the establishment phase. As evident in Figure 6.1, few of the families participating in these programmes continued their tree plantations through the harvesting phase. In one of the studied reforestation programmes in the Brazilian state of Pará – typical of the region – the smallholders received seedlings of forest species to be planted in degraded areas within their properties. However, only 30% of the initially participating families planted the received seedlings, and only 3% invested in the suggested treatments to ensure the good development of the trees, such as weeding. Consequently, nearly all the plantations visited in the field showed poor performance. Based on our observations, we estimate that only one of every 100 producers who initially participated in plantation programmes was able to produce and sell the originally envisaged products.

Obviously, there are problems that limit the local viability of these types of development initiatives, resulting in few families being able and willing to invest the effort and resources necessary for successful adoption. Regarding the above-described example of programmes promoting tree plantations, we found three main difficulties (Pokorny et al., 2010a; Hoch et al., 2012): (1) the long stages of production in comparison with the annual cultivation of agricultural crops; (2) the accumulation of risks over time in terms of damage and destruction by fire, drought, animals, infestations, floods and robbery; and (3) the difficulty involved in product commercialization, considering the high costs of transport and the lack of attractive markets. Tellingly, regarding commercialization, we found that, in the majority of the case studies, the markets that were initially expected for the products of interest simply had not developed over time. Withdrawal was also observed as a consequence of the price fluctuations typical of many agricultural products. The fall in prices of certain crops led to their abandonment, whereas the increasing prices of other crops created a dynamic wherein families often massively extended their production in an attempt to benefit from this opportunity (see Chapter 2). During such high-price periods, the smallholders also tended to ignore designated marketing channels (often cooperatives) and started to sell their products directly to traders. In addition to seriously affecting existing contractual obligations or the operation of rural marketing cooperatives, this practice sometimes led to the complete loss of collective achievements, for example, the withdrawal of producers from successfully operating agroforestry systems established with considerable effort over several years due to the high price of Brazil nuts in Bolivia (Hoch, 2009) or emerging markets for oil palm in Peru (März, 2008).

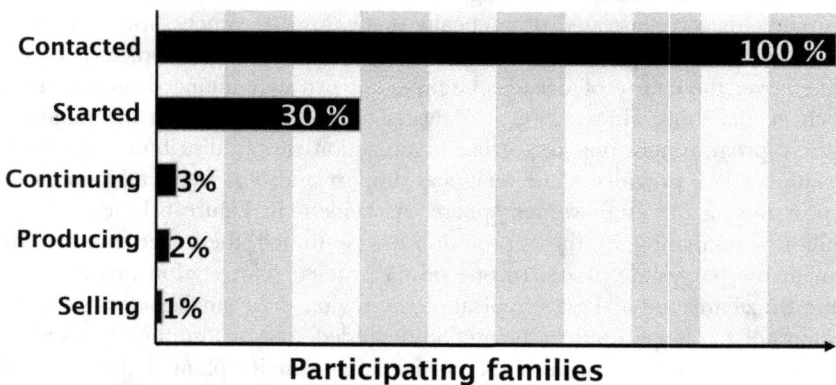

Figure 6.1   Proportion of forestry plantation adoption by surveyed smallholders in development programmes at each stage of production (adapted from Hoch, 2009)

Another indicator of the eventual incompatibility between the proposals and support strategies offered by the development organizations and the reality of the smallholders was the fact that we did not find any example of spontaneous replication of the proposed packages in the studied areas occurring independently of the promoted technical-managerial model and accompanying support. In fact, outside of the aegis of the projects and programmes, only several specific technologies, characterized by simplicity and adaptability to local conditions, were isolated from the packages and adopted by the families. Consequently, success stories related to development projects in the Amazon generally remained as isolated examples (Pokorny and Johnson 2008b).

Another example of the incompatibility of external efforts with the local realities was found in respect to the massive attempt by international environmental organizations to push forward FSC certification at the smallholder level to promote Sustainable Forest Management while alleviating poverty (Medina and Pokorny, 2008, 2011). Benneker (2008) analyzed experiments with FSC certification for the 'chiquitanos', an indigenous group in Bolivia, and demonstrated that the eventual price gains in the market fell far short of compensating for the related costs and efforts. This lack of financial viability was confirmed in all the relevant ForLive case studies. At a more general level, we found that the process of certification systematically favoured companies and other more capitalized actors due to their greater capacity to fulfil the technical and bureaucratic requirements (Pokorny and Phillip, 2008; Pokorny et al., 2012). Even when communities with robust external support achieved certification, the communities had great difficulty in maintaining the certification status after the supporting organization withdrew financing, not only for financial reasons, but also because several certification norms were in conflict with the local demands and practices of land and forest use – for example, when hunting was prohibited or severely restricted, or when the establishment of agricultural fields in designated forest management areas was forbidden.

The rapidly increasing number of REDD+ initiatives paying smallholders for environmental services with the goal of avoiding and/or reducing deforestation (Engel et al., 2008), which follows the conventional development approach of implementing externally designed management schemes at the local level, was also fraught with the risk of aggravating many incompatibilities (Pokorny et al., 2012). The government of the Amazonas State in Brazil, for example, paid a defined amount to families in the programme areas for their agreement to refrain from creating new agricultural fields in the forest and to stop the informal use of timber (Sato, 2010). However, in addition to the fact that the payments reaching each family were marginal in comparison to total revenues (and needs), the families could potentially be faced with the problem of declining productivity in their agricultural fields. The declining productivity results from agreements that force them to cultivate the same area consecutively over longer periods of time instead of following their traditional shifting cultivation schemes, in which they burn small patches of the forest to establish new fields that are then cultivated over a shorter time and subsequently left fallow to allow natural forest succession to regenerate the soil.

Additionally, in the case studies characterized by more long-term and intensive support – as noted, mainly initiatives with NGOs promoting agroforestry systems – the overall success rate was significantly lower than expected, although a larger proportion of participating producers maintained the promoted systems compared with other projects. For example, in one of the most successful experiments that we analyzed, a Bolivian NGO had promoted agroforestry systems with cupuazu (*Theobroma grandiflorum*) for more than 15 years but only 150 of the approximately 1,000 families supported during this period continued with the proposed management scheme (IPHAE, 2007). In this case study, one important factor that limited the successful adoption of the proposed system by an even larger number of families was the significant amount of time required to establish the systems. As evidenced by the comments of one of the successful farmers, 'I worked for the first three years during the day on my traditional agriculture to sustain my family and at night on the implementation of the new system. ... Only recently, after seven years of hard work and due to the support of IPHAE [the NGO], do the [cupuazu] trees generate a constant income'. In our case studies, we found that the majority of smallholders had neither the interest nor the capacity to overcome such an immense time/resource barrier (Hoch et al., 2009).

Creation of dependence

As highlighted above, the vast majority of technical models promoted in the development initiatives analyzed by ForLive insufficiently considered the local production schemes and suffered from a lack of compatibility with the necessities and capacities of the smallholders and thus with the environmental and socioeconomic realities of the families. Low productivity per hectare, high capital requirements and competitive disadvantages compared with professional companies seriously affected the long-term viability of these initiatives (Pokorny et al., 2012). In fact, nearly all the observed initiatives, contrary to the initial expectations of the supporting organizations, remained heavily reliant on intensive external support, even after several years. The organizational and financial capacities of most of the families were simply not sufficient to maintain the proposed technical and organizational models on their own. Consequently, these development initiatives generally increased the dependence of the receiving families on external support, notwithstanding the fact that many families, especially those who received direct financial and material support, benefited from the projects. We found a drastic example of this problem in the analyzed 'Community Forestry' initiatives. All of the communities that we analyzed suffered from severe cash flow problems caused by high capital requirements for the logging operations, complex refinancing mechanisms often delaying the receipt of revenues and a need for cash income. Combined with a general lack of experience with and a limited conception of financial management, this situation provoked a notorious shortage of cash to pay for operating costs and system investments. An extreme example of this shortfall was the 'Community Forestry' initiative in the case study of PAE Ecuador in Acre (Brazil), where more than US$200,000 was needed every year to cover the costs of harvesting operations; however, after two years, the families had still not been paid for their timber due to unclear and changing institutional arrangements for processing and other problems encountered with the commercialization of the harvested timber. Although, in this example, most of the costs were paid by the supporting organizations, this case demonstrates the magnitude of the problem. Even in the initiatives that generated profits, the families generally failed to accumulate a sufficient capital stock to guarantee the necessary liquid capital for the following years' operations.

In the case studies, we found that several development projects even negatively affected the livelihood basis of some families. In general, for these families, the time-consuming collaboration with development initiatives necessarily implied the often significant reallocation of local labour to the externally induced activities (Hoch, 2009). In some case studies, this shift had negative effects on the performance of traditional activities, particularly farming practices, which usually constituted the central pillar of local livelihoods (Vos et al., 2009). In extreme cases, this reallocation of limited resources even undermined the core aspects of local livelihood strategies and thus increased the vulnerability of these families to external threats. For example, we found that, when families were massively supported by NGOs, their traditional production schemes were replaced by the end of the project but they were not able to continue with the adopted production system on their own without external support.

One of the most critical findings of the case studies was the challenge of commercialization of the products generated within the new production systems at a price that compensated for the typically higher costs of production and increased risk (Hoch et al., 2012). In the development initiatives dealing with the production of timber, for example, most initiatives were oriented towards national and international markets in search of higher prices compared with local markets. However, this focus rarely allowed for the supported local groups to successfully establish stable direct relationships with the buyers. For agricultural products, the problem in the study areas was that the markets were mostly controlled by cartels of large companies and influential family clans. This situation was further aggravated by the increasing supply from strongly subsidized large-scale agriculture in traditional growing and industrialized regions and countries (Pokorny et al., 2012). The determining market mechanisms, even at the local level, left little room for independent commercialization by small-scale operators (Kern, 2012). This was particularly the case for classic export markets such as timber and Brazil nuts (*Bertholletia excelsa*) (especially in Bolivia) as well as for the lucrative emerging markets for oil palm (*Elaeis guineensis*) (in Peru and Ecuador) and açaí (*Euterpe oleracea*) (in Brazil). Thus, access to attractive markets mainly depended on the continued mediation of NGOs, which were typically located far away, often even outside the country (Scherr et al., 2001).

In view of the challenge of providing continuous financing, as well as support for the maintenance of machines, payment for consumables, replacement of equipment, updating of communication and information systems and fostering of contacts and networks relevant for commercialization, in several cases, the families or their representative organizations began to look for opportunities to continue cooperation with the same organization and/or with other organizations in new projects after the old project ended. Often, the collaborating NGO itself applied for new funds to maintain the initiative. In fact, fundraising was one of the principal activities of successful NGOs, which naturally attempted to maintain their teams and guarantee appropriate working conditions. In several cases, however, we observed the paradox that the NGOs' efforts to maintain their own functionality required more attention and more resources than the actual work with the target groups. In several case studies, we also found specific families who included the search for new opportunities for cooperation with external organizations as an essential component of their livelihood strategy (Medina et al., 2009a, 2009b). These families were capable of establishing themselves as a focal point for NGOs in their continuous search for adequate local partners for their projects. In very few cases, the smallholders or, more likely, their representative organization, managed to establish a more direct contact with possible donors as a basis for the development of their own project proposals. However, even in these cases, the donors still targeted the continuation of the new production models. In the case of 'Community Forestry' projects, however, the most common strategy was for the communities to continue with logging operations after the end of external support and to search for partnerships with commercial enterprises, which then took responsibility for most or all of the operations, including the commercialization of the products.

There were a few specific cases in which NGOs opted for a strategy of more continual support instead of conducting temporally limited projects of three to five years with changing local partners and donors. These NGOs naturally had more stable partnerships, with spatially concentrated efforts, and they provided intensive support for their selected local partners. Often, their portfolios included the establishment of training centres, manufacturing plants for the further processing of local products and associations for the organization of production and commercialization. These organizations also generally had a stronger political presence in local and regional discussion forums and decision-making mechanisms. Although all the observed NGOs in this category invested significantly in the inclusion of local people in their organization and explicitly promoted self-organization in the supported communities, they also showed a tendency to intervene strongly in the decisions of the families, thereby becoming a dominating, sometimes even paternalistic, influence. This situation often created local dependence while diminishing the abilities of local families to follow their own practices (Medina et al, 2009b). In an extreme example of this dynamic, we found in Acre, Brazil, that agencies established by the state government to provide logistical and technical support to communities engaged in timber management actually took full control of the timber production in local communities (Medina and Pokorny, 2008, 2011; Hoch, 2009).

In summary, the discussions with development experts showed that the ongoing repetition of the traditional role of professionals as the developers of solutions may seriously affect the adaptive capacities of smallholders in the long term and, in turn, limit local capacity to investigate problems and define solutions that could lead to the adoption of better practices for the more effective use of natural resources and local benefits (Porro et al., 2008, Medina et al., 2009a, 2009b).

Impact on the social system

Undoubtedly, the numerous development initiatives of governments, domestic society and international organizations helped to disrupt the historic system of feudalism and dependence prevalent throughout the Amazon (see Chapter 4). This dynamic had been based on unbalanced relationships between small-holders and powerful actors such as landlords, timber companies and traders, who constituted the economic and political elites of the region (Sabogal et al., 2008b; Pokorny et al., 2012). As noted by Larson and Ribot (2004), governments have increasingly recognized traditional rights of access to land and forests and started to establish more democratic and socially inclusive governance systems. The field research and stakeholder consultations performed in ForLive confirmed the observations of Humphries and Kainer (2006) and Donovan et al. (2008) that many families who participated in development initiatives benefited from direct or indirect transfers of funding, the provision of equipment and materials, training opportunities and improved market access; additionally, these families discovered, at least temporarily, new income opportunities. They were able to strengthen their formal organization and gain managerial capacities, particularly in the financial and information spheres (Donovan et al., 2008). Thus, some of the families in the rural Amazon managed to improve their living conditions, sometimes significantly, and learnt to interact more intensively with each other and to develop closer working relationships with external actors. Many of the representatives of development organizations interviewed during the project highlighted these indirect effects as a crucial contribution to the enhanced ability of smallholders to face challenges and to take advantage of emerging economic opportunities and public policies.

However, the development initiatives typically demanded that the families in their projects learn new technical, organizational and managerial skills, engage with new markets and comply with new regulations and institutions (Sabogal et al., 2008a; Pokorny et al., 2010b). Consequently, the families were obliged to perform tasks and comply with functions that often did not exist before (Medina et al., 2009a, 2009b). The analysis of the case studies revealed that this strategy affected daily routines and power relationships within communities and, like any intervention in a social system, provoked unpredicted and negative consequences in addition to the intended positive effects described above. In particular, they affected the ability of the smallholders to play a leading role in the management of their resources with their own social identity (Porro et al., 2008; Medina et al., 2009a, 2009b; De Koning, 2011). Based on the conclusions drawn during a sequence of events held to present the findings of ForLive regarding the effects of development initiatives on the social system, the analyzed case studies can be classified into five broader scenarios: (1) the *success* scenario, wherein the supporting organizations managed to replace local structures with the propagated, externally defined model of production and organization; (2) the *elite capture* scenario wherein elites took advantage of their social position for individual benefit; (3) the *parallel institutions* scenario wherein the project induced the establishment of new institutional structures working in parallel with the traditionally existing local institutions; (4) the scenario of *social separation* between individuals successfully working within the projects and the remaining community members; and finally, (5) a scenario of *indifference*, wherein the external initiative had no significant effect on the families or their social system.

In the *success* scenario, extensively described by Stoian and Donovan (2004) and Macqueen (2012), the smallholders successfully adopt the socio-productive model suggested by the development organizations. In the analyzed case studies, these external proposals primarily aimed to substitute local production schemes with models based on the logic of enterprises, i.e. efficiency and competitiveness. In practice, however, we found almost no examples of this type of complete transformation. In fact, even in initiatives that received massive investment from the development organizations in the development of more professional structures for production, commercialization and organization over a longer time period, the vast majority of the families showed a clear tendency to maintain crucial features of their own local socio-productive model. In contrast to the proposed enterprise model, for example, they put a greater value on leisure time (natural pauses and intervals in the day) and focused on maximizing the possibility of compensation for labour rather than profit (Medina and Pokorny, 2008, 2011). The fact that the initial costs of a project, e.g. for the purchase of machinery, administrative assistance and training, were often paid by the support organization also induced a certain carelessness regarding the logic of capital return. In contrast to the common business model of enterprises, within community enterprises, participating families operated in more horizontal structures and, consequently, exerted less pressure on the persons involved in the production process to follow determined tasks, guidelines and timelines. Finally, nearly all families involved in the initiative showed a high interest in continuing their traditional production activities and practices in parallel. Consequently, the productivity levels of the community initiatives were in general lower than in conventionally operated companies (see also Chapter 2), which seriously affected the competitiveness of smallholders' products in the free market (Pokorny et al., 2012). In fact, even those families who managed to successfully adopt the technologies and organizational proposals continued to rely on external support in the long term (Pokorny et al., 2010b).

In the *elite capture* scenario, the opportunities associated with collaboration with the external organizations where mostly captured by a few wealthier families who were able to participate more easily because they had more capacity and experience in communication and had some capital available (Hoch, 2009; Hoch et al., 2009). We found that this scenario reflected the sometimes well-differentiated local hierarchies but was often also induced or at least aggravated by the way in which international cooperation functions. For example, in the case of support through NGOs – which generally depended on external funding – the pressure by donors to generate tangible results in relatively short time periods meant that implementation and support on the ground was often focused on the most accessible and promising families and communities. There was a clear tendency in many organizations to concentrate on the families and communities known to be capable, interested and easily accessible in terms of communication, for practical and logistical reasons (Depsinszki, 2007). In the discussions and ForLive workshops, the strategy of selecting the most capable families with the most resources and largest labour force was justified by the expectation that, once these families had managed to successfully establish working practical examples of the promoted production schemes, the other families would follow. However, in practice, the poorer families were far from having the opportunity to replicate the trial schemes, as they lacked an adequate level of external support, which, instead of being stronger than the support given to the local elites, was often already terminated by the end of the project. As such, initially existing social inequalities increased as the privileged families accumulated more resources while the poorer families fell even farther behind. Frequently, this outcome provoked conflicts between the families. Only in cases where the local leaders were highly conscious of their social responsibility was this consequence avoided, at least in part.

In contrast to the above *elite capture* scenario, in several case studies, the development organizations worked exclusively with people of lower social status. In some of these cases, the local elites declined to collaborate because they did not buy into the benefits (especially individual benefits) promised by the project or because they were concerned that the initiative might undermine their dominant and favourable position. In other cases, the external organization explicitly worked with young people or focused on strengthening socially marginalized groups. In particular, for many development organizations, the empowerment of women was part of their mission statement to alleviate discrimination common in the region or to unlock the large potential and capacity of women to perform family welfare activities. Often, intervention recipients were supported on an individual basis with intensive training, access to information and the opportunity to travel; this gave them a chance to establish new contacts and experience new realities outside the context of their community, which significantly improved their personal performance and well-being. In several cases, this strong individual progress contributed to the establishment of new power structures operating in parallel with the traditional ones. In the *parallel institutions* scenario, this duality created a situation that occasionally provoked conflicts with the former authorities – whether in a more general confrontation of the traditional way of life against new models of thinking and acting, or more specifically with respect to the mode of distributing costs and benefits, the right to use common resources and the responsibilities and mechanisms for decision-making. In outstanding examples, these processes contributed to a better integration of socially marginalized groups and helped to overcome historical social barriers. In other cases, however, we found that the emerging conflicts weakened the actual socio-productive capital of the community. In particular, several of the individually supported young people became frustrated by the limitations of their social reality with respect to utilising their newly gained skills and perspectives. In more than one case, these individuals decided to abandon their community and migrate to the local urban centres, to bigger cities or abroad, giving rise to a scenario in which *successful individuals left their original social system*.

The great majority of the observed initiatives, however, resulted in surprisingly few changes in the local context (Gasche, 2004; Hoch et al., 2009). Often, the programmes and projects lacked the potential to reach a large number of families due to very limited resources or immense formal bureaucratic requirements, or simply because the project did not include strategies designed to reach a wider range of families. Another important reason for the lack of effect can be attributed to the temporary and isolated nature of many of the external organizations' initiatives, which were generally marked by single actions without much continuity that typically lasted no longer than three years. Furthermore, the organizations' contacts with the families were often limited to only a few field visits in which the technicians did not even spend a night in the community (Pokorny and Johnson, 2008b). Finally, many projects also suffered from a lack of support by the powerful individuals of a community or smallholder group and were therefore forced to collaborate with people, often young, who did not have responsibility but were available at the time to assist with meetings, training activities and local activities. In these cases, the other families did not consider their recommendations, and things continued as before.

In our case studies, the vast majority of observed initiatives designed to support families did not adequately account for the existing organizational schemes within the communities. There was a gap between the locally developed structures, mechanisms and practices and those required and promoted by the development projects. As a result, in the local development initiatives pushed by external agents, we generally found that the original structures coexisted with the more formal newly introduced structures. De Koning (2011), in her PhD research on six traditional and indigenous communities in Bolivia and Ecuador, identified three local responses in addressing these external stimuli: (1) in some cases, the families partially adopted the external models suggested by the supporting organization by recombining their local institutions with the new ones, and vice versa, so that the institutions often became multipurpose ('aggregation'); or (2) the families began to change the intention of the external proposal by constructing an – often virtual – context in which these proposals were rejected or altered in accord with the families' tradition ('alteration'). Often, the initially rejected proposals were adopted over time. Alternatively; (3) the families rigorously rejected the proposals as being incompatible with their traditions and realities ('articulation'). This drastic opposition, however, often created barriers for any type of externally proposed innovation, regardless of its utility. Over time, this lack of flexibility and openness strained the traditional institutions.

As described previously, this contrasting parallelism of locally and externally defined norms and mechanisms, depending on the specific framework for collaboration, occasionally strengthened the traditional elites or promoted the establishment of new elites and structures that were not necessarily respected or accepted. The resulting conflicts and insecurity showed the potential to advance the local organization of families, but also often weakened their social capital, thereby increasing their vulnerability, especially their dependence on external support. This problem was even more relevant when the development organizations systematically overestimated the potential of their own proposals to consistently generate local benefits in comparison with the existing local productive and social activities. Consequently, the existing local potential for families to induce their own processes of social change and overcome historically unjust situations and structures (see Chapter 4) was rarely mobilized by the supporting organizations. Thus, the positive outcomes of overcoming injustices and political insurgency typically appeared as a by-product of the initiatives or as mere coincidence. In contrast, the majority of the supporting organizations consciously avoided any conflicts with formal and informal authorities to avoid threatening the technical success of their projects or, particularly with respect to international organizations, to avoid compromising their delicate legal status in a foreign environment.

Outside the projects

Perhaps the most worrisome observation with regard to the various tree- and forest-related development initiatives analyzed by ForLive was not the difficulties involved with continuing the projects or the possible undesired social effects caused by the interventions, but rather the lack of replication. In almost none of the cases studied did the families who were not direct participants in the initiatives assimilate the proposals disseminated by the development organizations. Apparently, without massive external support, the smallholders were not able to comply with the financial and technical requirements associated with the adoption of the external proposals or they were simply not interested because of a lack of attractiveness. In several initiatives, families sought to be incorporated into the project to receive the offered external support. Very rarely, other organizations assimilated the agenda of temporally limited development initiatives that were perceived as successful to aid their continuation and propagation. In this case, governmental organizations mainly adopted the initiatives of internationally financed NGOs. However, these cases also suffered from major replication problems due to a lack of human and financial resources in the governmental agencies. In general, governments faced immense difficulties in properly translating isolated examples of success from projects into public policy. Nevertheless, certain policies and governmental efforts, particularly those related to the regulation of land tenure, credit programmes, better control of natural resources and the demarcation of concessions, had an influence on smallholders.

Regulation of land tenure

Nowadays, there is a broad consensus that securing smallholders' rights to customarily owned land and resources is a prerequisite for their effective use of forests and sustainable development (Larson et al., 2010, Sikor and Stahl, 2011); however, only during the last two decades have the territories of indigenous groups and other long-term residents been formally recognized. In the majority of these cases, the tenure rights were collective, and the transferred areas received protection status. These areas proved to provide effective protection against deforestation, especially when public control was mobilized (Nelson and Chomitz, 2011; Porter-Bolland et al., 2012), but the protected area status had mixed outcomes for the residents (Ehringhaus, 2005). Resident families were protected from outsiders trying to claim land or forests, but they were also forced to observe a cumbersome administrative process to use the same resources commercially. Moreover, the threats to these territories continued, including illegal invasion by logging companies and gold seekers (Tacconi, 2007), expropriation for the purpose of mining or the construction of dams and other infrastructure projects (IDB, 2006), and the selling of timber rights to logging companies, oftentimes benefiting a few powerful local leaders or families (Cano, 2012).

Governments also invested in the regularization of the tenure situation in the highly conflictive colonized areas. For instance, the governments of Brazil, Bolivia and Ecuador recently inaugurated ambitious programmes to certify land ownership by small colonists. However, in all our study areas, we observed a marked inability of public organizations to address the complexity of the shifting situation (Larson et al., 2008; Wagner, 2008; Pokorny et al., in press b). Often, the existing efforts of the authorities to legally acknowledge the land tenure of the smallholders only managed to identify a wide array of difficulties but did not resolve them. In fact, the majority of governmental agencies established to manage the legalization process have failed to demarcate smallholders' properties due to their limited capacities. However, even in instances when this difficulty was overcome and the land boundaries were defined, the bureaucratic processes for ownership certification were complex and required a long time, generally no less than ten years. Accordingly, the land regularization process in all study areas advanced slowly, leaving an enduring risk of conflicts between the local families and external actor groups (Pacheco et al., 2011). To a certain degree, the delivery of formal land tenure to the settler families accelerated the emergence of new rural-urban networks and more extensive management, with the abandonment of rural properties (Padoch et al., 2008). Settling in the agricultural frontier and selling land after a period of time has become a part of the livelihood strategies of many settlers (Hecht, 2011). Although owning individual legal property rights facilitates this dynamic, land is also sold without formal land titles, particularly after the forest has been slashed (Pokorny, in press a). Only settlers with holdings on fertile soils and better market access showed a tendency to stay in the same locations and create environmentally stable cultivated landscapes (see Chapter 3). The future long-term success of such programmes will therefore depend on the capacity of the technical teams to understand the specific characteristics and needs of the families and communities (Wagner, 2008) and political continuity; however, success in such endeavours is not typical in Latin America, where governments are known for discontinuity, abrupt changes and political upheaval.

Credit programmes

As smallholders generally suffer from a lack of capital, the provision of credit is a very popular policy in development initiatives (Pokorny and Johnson, 2008b). Credit programmes are typically executed through the banking system, which was previously established in many rural towns throughout the Amazon and therefore reaches many smallholders. However, public credit programmes in the Amazon mostly target actors who are better off and are often already engaged in commodity markets (Pokorny et al., in press b). Thus, the majority of credit programmes cater to formalized actor groups that are often led by wealthier families or even by largeholders with political influence, and, in our case studies, they tended to be utilized by families who were initially better off.

In the case of credit programmes developed to support smallholders, we generally observed a disappointingly low level of success in terms of outreach. Although the public credit programmes have become more flexible over the last few years, individual families still have little chance of overcoming the bureaucracy and fulfilling the formal requirements to obtain a loan. Often, the formal requirements of these programmes impeded many families, especially the poorest, in their ability to access the offered credit because they did not have the necessary documentation, the capacity to fill out the forms or the resources needed to present the documents to the banks or lenders, which were typically located in larger cities. In many cases, neither the private nor the public banks accepted small farm properties as collateral for loans, because land tenure was not formalized, the value of the land was too low or the families lived on commonly or publicly owned land in which the individual smallholding could not be impounded. The few programmes explicitly designed for poor rural dwellers are often not meant for individuals but require membership in a formal organization such as an association or cooperative. This situation has resulted in the emergence of a huge number of formal associations throughout the Amazon region that were created to access support programmes established by governments, banks and NGOs. The majority of these associations, however, function poorly and often disband after receiving the targeted support.

Another problem with the credit programmes observed in the study regions was that they were mostly tied to the transfer of technology packages (Pokorny and Johnson, 2008b) that had little utility for smallholders or they promoted production systems that were inadequate for the environmental conditions of the region, most prominently in the case of the subsidies for cattle ranching in Brazil (Arima et al., 2002; Pacheco, 2009). This problem exists partly because the majority of credit programmes are geared towards regions with older agricultural traditions, stronger infrastructure and higher degrees of market integration. Consequently, these programmes are not well adapted to the realities of Amazonian smallholders, who generally have a lower level of mechanization and typically produce for both markets and their own consumption. For the same reason, available credit is nearly exclusively limited to settlers, whereas indigenous groups and other long-time residents are rarely targeted.

Therefore, even with the successful implementation of these programmes, there was widespread abandonment or failure of the financed activities. Generally, the loans provided were designed as packages with limited flexibility to accommodate local conditions or the needs of the smallholder. Another difficulty emerged from the fact that the public systems for technical assistance were tightly – often even exclusively – linked to the preparation and support of projects financed by these credit programmes, and they did not adequately address the needs of the small-holders. Many times, the technicians did not visit the area of the smallholders before the execution of the project and had little capacity to support them during its execution (Pokorny and Johnson, 2008b). Interestingly, we found that many families took advantage of this situation by ignoring the rigid rules and using the funds for their most urgent needs or personal preferences. With respect to production, the majority of families invested in the expansion of agriculture, for example by buying cattle or new land for cattle ranching (Herrera, 2011). In Medicilândia, we found families who systematically exploited these opportunities to access funding: all members of the family applied in succession for available public credit programmes. However, only some of the recipients repaid their loans. Thus, successful farmers benefiting from loan programmes occasionally increased the pressure on poorer families, who were often forced to leave their land, with the consequence that their rather diversified small-scale production systems disappeared with them.

Control of natural resource use

Efforts to revise the legal and institutional frameworks to more effectively regulate and control the commercial use of natural resources deeply affected the realities of smallholders. As described in Chapter 5, the new regulations did not sufficiently account for the huge diversity of local realities and natural resource management schemes, but instead indirectly favoured the application of standardized forest management schemes that depended on greater inputs and technical requirements that were not feasible for smallholders without massive external support (Sabogal et al., 2008b; Medina and Pokorny, 2008, 2011). In fact, in place of the regulatory efforts ensuring smallholders' access and rights to their resources, the new legal prerequisites, in combination with a notorious lack of institutional capacity for effective law enforcement, ultimately further complicated the situation of the vast majority of families, creating more conflicts and opening the door to corruption and impunity. However, past experience showed that regulations prohibiting people from entering forests, hunting animals, cutting trees for timber, fruits or palm hearts and cutting patches to make swiddens have largely been ineffective (Garnett et al., 2007).

Actors with more resources, especially timber companies, systematically utilized the legal 'vacuum' or the gaps created by the inconsistencies and deficiencies in the regulations (Weigelt, 2011). As a result of the complications and changes in use required by the legal-institutional framework, smallholders were discouraged from continuing their traditional modes of forest use. Thus, the reforms undermined existing local institutional arrangements to control the use and management of natural resources, particularly in the case of traditional communities (Medina et al., 2009a, 2009b); instead, they often legitimized the ransacking of the forests in traditional lands. The requirement for formal land titles and management plans signed by forest engineers, the need to create legally and formally registered organizations for the sale of products and, in several countries, the prohibition against using chainsaws to process timber, led the smallholders, who previously worked informally, to work illegally. This 'illegalization' of the smallholders' economic activities, along with improved control mechanisms, further weakened the smallholders' negotiating position in their interactions with timber enterprises, intermediaries, traders and customers, forcing prices down or obliging them to bribe local authorities to obtain the necessary authorizations (Thomas, 2008). In the case studies in Peru, for example, we found that, after the legal reforms, the intermediaries and the buyers of forest products paid lower prices to the families, arguing that the products were of illegal origin. In Porto de Moz, Brazil, some of the families even stopped – at least temporarily – using their forests due to the fear of entering into conflict with the authorities and being penalized for their activities. In particular, the traditional communities in areas endowed with legal protection status, although they benefited from the improved security of land tenure, faced difficulties in continuing their traditional activities in a legal way because the authorities generally demanded management plans, formal registration and integration in bureaucratic procedures (Weigelt, 2011).

As described by the World Bank (2008), the advances in the regulatory framework systematically strengthened the more powerful commercial actors, who have proven to be much better prepared to cope with the stricter legal-institutional framework and market requirements; these capitalized actors complied with the framework because they had access to capital and information, qualifications and institutional relationships or because they were able to circumvent the new requirements more easily (Thomas, 2008). These 'large' actors also had less difficulty in applying for credit or utilizing the opportunities offered by international and national markets. In most situations, the private sector was able to take advantage of its competitive advantages (Pokorny et al., 2012). In the case of forest management, the timber companies, conscious of the difficulties that communities experienced in fulfilling the new legal, technical and management requirements, revived efforts to access the forests owned by smallholders and communities, who often had no realistic chance of achieving legal status for their forest operations. In fact, in all the study areas, we found that timber companies entered into increasing numbers of negotiations with the communities to exploit locally owned forests, as indicated, for example, by the enormous rise in the number of forest management plans in smallholder forests in Brazil managed by timber enterprises, which went from just a few to several thousand. As described by Lima et al. (2003), in these agreements, the companies assume responsibility for complying with the legal requirements such as the elaboration of forest management plans, providing technical support and carrying out field operations, and they pay the families for the products, sometimes also advancing payments to cover local investments in time and material. However, the in-depth analysis of such a case study in Rondônia (Pantoja, 2008) indicates that such agreements induce predatory behaviour by the commercial logging operations, with the benefits for the smallholders remaining little better than under the former illegal schemes. Like Sabogal et al. (2007), we found that the large majority of companies paid low prices for the timber that they obtained from the smallholders' forests (arguing that they bore the costs of legalization) and created extensive damage in the forests, thus leaving the smallholders with degraded forests and with little possibility of legally accessing the remaining products until the end of the cutting cycle defined in the management plan. In the Bolivian case studies, we encountered several communities that had previously signed contracts with companies for 20–50 years with fixed prices that were considerably lower than the actual market value. Thus, the legal reforms complicated the situation of many families and accelerated the predatory use of the forests by timber companies.

Concessions

The demarcation of concessions in public forests for the commercial use of timber did not translate into benefits for smallholders in rural zones. In the worst cases, the designated concessions included areas of traditional use and sometimes even settlements – a situation that predictably led to conflict, which, in the vast majority of cases, was 'resolved' in favour of the large concessionaires, as the smallholders had no clear legal status and were not able to verify their rights, or because their rights were simply ignored. Even in cases in which the authorities attempted to consider the traditionally existing land tenure situation and rights to access natural resources in the demarcation of concessions, as in Bolivia and Brazil, the families living inside or next to forest concessions were seriously constrained in continuing their traditional forest uses (Benneker, 2008; Pokorny et al., in press a).

Smallholders were effectively excluded from the bidding process due to technical, financial and bureaucratic barriers. Because a medium-sized sawmill usually needs an area of at least 30,000 hectares during a cutting cycle of 20–40 years (Pokorny and Steinbrenner, 2005), a forest concessionaire must invest in heavy machinery and equipment and comply with strict regulations. This makes it nearly impossible for smallholders to successfully bid in public auctions. Even medium-sized logging companies find it difficult to mobilize the required capital and professional capacity, although several state governments actively promote their involvement in the forestry sector (Brazil's National Press, 2005).

Only in very few cases did smallholder communities obtain concession rights. For example, in the Ambê initiative in the Tapajós National Park (Brazil), local river communities obtained exclusive rights to manage approximately 25,000 ha of public forests (Medina and Pokorny, 2008). Where communities have succeeded in establishing legal property rights, as in the case of indigenous and extractive reserves, they also obtained the exclusive rights to commercial use of forest products; however, they had to prepare forest management plans to obtain these rights (Chirif and Garcia Herrero, 2007). In Bolivia, public forest concessions were allocated to groups of local dwellers called Agrupaciones Sociales del Lugar (ASL; Ruiz, 2005). Peru and Bolivia also adopted concessions for non-timber forest products, particularly Brazil nuts (Masias, 2011). In Brazil, the government also implemented Sustainable Development Projects (Projetos de Desenvolvimento Sustentável; PDS), which are settlements based on the management of collectively owned forests. However, in all these cases, smallholders faced insurmountable difficulties in complying with the administrative and technical rules, and in the few successful cases, they relied heavily on massive external support (Medina and Pokorny, 2008, 2011; Pacheco et al., 2010).

Finally, it also became evident that, in the study areas, families in rural areas benefited little from the jobs generated by the concessionaires in the forest areas because the majority of the companies brought their own staff to work in the harvesting operations or because the work conditions were not compatible with the needs and demands of local families. Thus, when forest concessions were granted, employment was mostly generated in the saw mills, which are usually located near urban areas. This, in turn, contributed to the problem of migration and urbanization.

In summary, our studies revealed a paradoxical situation: after more than 20 years of efforts to promote 'Community Forestry' as a promising alternative to the exploitation of communities by the timber companies, these very efforts have consolidated a situation wherein the companies were able to act in the same destructive way, with only one significant difference: nowadays, many companies pursue their activities legally and with the support of not only the government, but also international organizations that perceive the inclusion of the private sector as the most effective means to guarantee professional, high-quality management of forests in the Amazon (Masias, 2011; Pokorny et al., 2012).

Accelerating marginalization

As described above, the ForLive studies revealed a rather complex and ambivalent picture with respect to the consequences of the attempts of external organizations to improve and regularize the situation in the Amazon. The initiatives to support the rural poor, who generally live in difficult circumstances, encountered structural problems that often originated from the historical context of exploitation and oppression of the local populations (see Chapter 4). Moreover, many of the development organizations working with smallholders, especially those targeting traditional communities and indigenous groups living on land with a high proportion of remaining natural forest, had a predominantly environmental mission and were thus challenged by the attempt to combine environmental protection with the rational use of natural resources, generally, the only resource in the hands of the families with any economic value. Nevertheless, a series of initiatives working simultaneously at the local, national and international levels achieved significant advances for the local smallholders. That said, these efforts demonstrated many inconsistencies and were mostly marginal compared with what was achieved in other sectors and for other actor groups. In addition, a serious structural problem emerged in the fact that the ideas of modernization, trust in markets and belief in equal chances – which served as guiding principles for the vast majority of these initiatives for local development – are not necessarily compatible with Amazonian realities (Pokorny et al., 2012).

Simultaneously, for the vast majority of society, the integration and adaptation of rural populations into a globalized market, value chains, production schemes and societal structures is perceived as the only real possibility of improving the precarious situation of these families and enabling the development of the region. The families in the case studies understand that the construction of new roads and the possibility of commercializing their products in global markets are the most realistic option that they have to improve access to public services such as education, health, credit and technical assistance and to also avail themselves of opportunities for commercialization and consumption.

However, in the study areas, the rapid proliferation of this approach to development created a climate in which the smallholders exchange an increasing share of their lifestyles and practices related to natural resource management for models of production and organization developed and dominated by external actors. This was especially true in the case of indigenous and traditional communities, where we observed a process of erosion of local knowledge and capacities regarding traditional land-use practices (Porro et al., 2008). The new models perpetuated the same logic of global markets, which generally focuses on a limited number of commodity products and production systems that require significant investment and a greater economy of scale. These models also require more hierarchical systems of organization and business structures rooted in functional aspects that guarantee high productivity, efficiency and profits (Macqueen, 2012). Consequently, although these initiatives aimed for just and sustainable local development, in spite of the crucial benefits that they generated for many families in the region, overall they tended to accelerate the marginalization of smallholders because they consciously and unconsciously promoted the adaptation of families to the demands of global markets. This strategy – so strongly centred in a classical development model – necessarily raises many risks for the future: the danger of contributing to a process of deterioration of the large cultural diversity of rural populations, of forcing or stimulating the processes of migration and urbanization and consequently further accelerating the transformation of the forest on a large scale (Pokorny et al., 2012).

7 Barriers to equitable local development

The previous chapters have demonstrated that current efforts for rural development have tended to widely ignore the potential of smallholders to contribute to sound local development and instead consider poor local people more as a problem, as an obstacle to development or, at best, as the principal beneficiaries of development efforts driven by actor groups with the financial and human capacities to manage large-scale professional investments. In accordance with this view, nearly all the analyzed support initiatives targeting rural families were mainly concerned with the integration of smallholders into global commodity markets rather than with understanding the smallholders themselves as epicentres for local development. Consequently, the majority of the observed initiatives focused on the modernization of local systems for production and organization. In the previous chapter, we described how this approach, although it generated significant benefits for many rural families, necessarily contributes to the homogenization of the original diversity of social and productive systems in the region and thus results in the further marginalization of small producers, traditional communities and indigenous groups. This, in turn, accelerates the ongoing processes of deforestation and environmental degradation. Thus, this development perspective (unintentionally) risks reinforcing the barriers that impede sound development and the long-term improvement of the general situation of the vast majority of families living in the rural Amazon (Pokorny et al., 2012).

In order to realistically evaluate the potential of alternative development paths rooted in strengthening, rather than replacing, the existing capacities of smallholders' socio-productive systems as the basis for sound local development, it is important to be conscious of existing barriers and the possibility of overcoming them. With this goal in mind, this chapter discusses the principal factors that limit or even impede development programs in the Amazon that work with the cultures and capacities of the local populations and their environmental resources. We can thus distinguish the barriers created by external actors from those caused by the smallholders themselves, which generally result from these external obstacles.

Barriers in the external framework

In a world where the majority of nations and people remain oriented towards variations of the classical model of economic growth as outlined by Rostow (1960) half a century ago and practised by the industrialized countries of North America, Europe and Asia, and in light of the economic success of so-called BRIC countries (countries at similar stages of newly advanced economic development, in particular referring to Brazil, Russia, India and China) (O'Neill, 2001), we expect many barriers to the creation and implementation of an alternative vision for development based on the capacities and realities of smallholders, who are universally excluded from global financial and commodity markets. Beyond the general difficulty of overcoming the traditional approaches to development, the reflections realized within ForLive revealed a number of more tangible barriers: (1) lack of respect for the smallholders and ignorance of their capacities; (2) insufficient knowledge of the smallholders' realities; (3) an overwhelming global economic dynamic that is beyond the control of policy makers and, somewhat related, the major interest of powerful economic actors in the resources of the region; (4) inconsistency within the policies of various sectors; (5) the state's weak ability to act in rural zones and an institutional framework defined by corruption and patronage; and finally, (6) unfavourable production conditions that make it difficult to identify financially attractive options for the smallholders. The following analysis discusses these obstacles in more detail.

Lack of respect for local capacities or 'thinking inside the box'

In the experiments studied by ForLive, it became obvious that the attitudes of external actors who deal directly or indirectly with smallholders represent one of the principal barriers to appreciating the value of smallholders' cultures and capacities. The prevalent discourses and the propagation of beliefs strongly linked to the classical model of economic growth within the rural education system and mass communication channels systematically fosters reproductive patterns of action instead of mobilizing the personalized knowledge needed to create new solutions (Polanyi, 1983; Flick, 2008). As a consequence, reflection remained within a box of fluctuating social structures and meanings produced for this information and information exchange (Deetz and Radford, 2008). Effectively, the large majority of political representatives, heads of development organizations and technicians contacted during the four years of the project held the viewpoint that the classical model of development through fast economic growth (and its assumptions about progress) is the best model available, widely ignoring the negative social and environmental effects or accepting them as the necessary costs of 'development'. Thus, the existence of alternatives was rarely considered by the relevant decision makers and development agents.

There were significant difficulties in terms of communication because the people working in public or private organizations for rural development, who have had several years of study in schools and universities, showed a marked resistance to accepting – or even considering – the possibility that the smallholders may know more about certain aspects of their socio-environmental reality than an expert, and that locals possess the capacities to plan viable action based on local decision-making processes (Pokorny and Johnson, 2008b). For instance, in many of the discussion events held within ForLive, it was difficult to even discuss whether and to what degree the production systems applied by smallholders, characterized by low inputs and family-based labour, might be an appropriate option in the local context. Typically, after some agreement on general statements about the 'importance of local knowledge' and the 'capacity of local people', the discussion rapidly turned its focus to opportunities to improve the 'technically and organisztionally deficient' local socio-productive systems. These assumptions led to conclusions regarding the 'magnitude of the challenge' of creating local capacity to make use of available knowledge and technologies and 'emerging market options'. It became obvious that the vast majority of policy makers, professionals and technicians held the opinion that smallholders, by definition, suffer from insufficient knowledge and skills to effectively manage their resources and that only a business plan based on expert knowledge and technological input can generate the outcomes needed to advance local development.

This attitude was deeply ingrained within the system and perpetuated as a normalized attitude by actors at all levels; consequently, these knowledge-management relationships made any 'thinking outside the box' challenging. This problem was confirmed by observations in hands-on field work: field workers were often so concerned with the transfer of technologies developed by experts (e.g. nurseries, seedlings, forest management plans and agroforestry systems) that they did not even consider the possibility that the families might be able to develop or contribute to the development of their own solutions to their problems (Pokorny and Johnson, 2008b; Flick, 2008). Opportunities to promote dialogue, knowledge exchange and communication among members of the smallholder communities to generate conditions of mutual learning were rarely utilized. Instead, the main thinking role was assumed by the technicians in a one-way dynamic that was rarely of use to the smallholders and that furthered their dependence on the technician for support.

This phenomenon was also apparent in the ForLive project's own attempts at an action research component (see Chapter 1), wherein the project assistants were asked to assume the role of facilitators promoting local research agendas rather than acting as experts (Pokorny, 2003). The professional identity of the majority of the project assistants as qualified experts in resource management made it nearly impossible for them to act as facilitators in local initiatives. Instead, they consistently tended to guide and explain how things should be performed, focusing in particular on local practices that, from their point of view, were techni-cally deficient and differed from what they had learnt in their formal education. In practice, even those technicians who explicitly stated that local practices may make sense and should be more intensively considered in local development efforts systematically sought deficiencies in local action rather than identifying existing positive aspects or critically analyzing the reasons behind the locally developed management methods. Thus, even in the ForLive action research component, the project assistants rapidly began to work on the transmission of technologies and modes of organization (Flick, 2008).

In the workshops and interviews, these difficulties in overcoming classical development thinking and accepting locals as partners for development rather than as clients of development aid was apparent at all levels, including researchers, NGOs, policy makers and donors (Pokorny and Johnson, 2008c). In fact, examples of local families effectively using the supporting organizations to bring forward their own ideas at a more strategic level were extremely rare (Medina et al., 2009b).

Lack of knowledge about smallholders' realities

One reason for the lack of consideration and respect for local cultures and capacities identified in the ForLive reflection events was that people working in governmental and non-governmental organizations disregarded the meaningfulness of local action because of their insufficient understanding of local realities. Nearly all the coordinators of the governmental and private development organizations contacted in the study areas were of middle-class origin and were educated in schools and universities located in the larger urban centres, often with no immediate contact with rural areas. Additionally, the majority of the organizations themselves were based in cities in order to benefit from better working conditions and to guarantee contact with decision makers and donors. Consequently, the personal and institutional realities of the people engaged in development issues differed significantly from those of their target groups; this engendered a profound lack of understanding of the smallholders' realities, which was exacerbated by the fact that people in management positions did not have time built into their roles to travel to rural areas and meet with the target groups. Furthermore, there was a lack of effective mechanisms for direct communication with the smallholders. Paradoxically, in spite of their distance from local families, it was NGOs and other intermediary organizations, rather than representatives elected by the smallholders themselves, that typically represented the positions of the smallholders in public debates (Medina, 2008). Consequently, almost all the development-related programmes, projects and infrastructure endeavours in the Amazon were proposed by people outside the region, with little participation from local people and insufficient connection to local realities. In fact, the economic and political foci of almost all the countries sharing territory in the Amazon were located in markedly different contexts and ecosystems: in Brazil, the economic and political centres are located in the South, whereas in the Andean countries, they are typically found in coastal areas or areas surrounding large cities. This literally 'external' origin of Amazon development initiatives is further exacerbated by the large number of specialists, politicians and businesses in industrialized countries who influence discourses and policies.

Astonishingly, even the technical staff working in the development organizations generally had little direct contact with the families in the field. The technicians typically did not stay very long with the families or in the communities, as they mainly worked in their offices. When they visited the field, many technicians interacted primarily with the local leaders or a few families who were previously accustomed to communication and collaboration. In order to improve contact with the communities, several of the larger NGOs trained local people to serve as development aides. However, these local people typically remained in inferior positions in the hierarchies of the organizations and therefore had little influence on decision-making processes. Most critical was the observation that the majority of the permanently employed experts and technicians who were most capable of communicating and exchanging ideas with the locals worked at the managerial level of these organizations, whereas the employees who worked directly with the communities were paid less and occupied their posts only for a short time. The employees in this latter group were deeply concerned with achieving the goals set by the projects through which they were paid, and they did not show strong interest or the capacity to reflect critically on how the local context operated and what possibilities might exist to adjust the initiative to local interests and capacities (Pokorny and Johnson, 2008c).

In the universities and research organizations active in the study areas, where much of the information that informs future development strategies and projects is generated, the situation seemed to be even more distant from the local reality. Generally, the professional training in the universities was not oriented towards the characteristics and capacities needed to interact with rural families and communities. In fact, the majority of the university courses lacked any type of option for students to become familiar with local realities, instead focusing merely on the generation of technical know-how (Santana et al., 2003). Overall, we found that the majority of students, despite their general interest in environmental and social issues, were not interested in careers that would imply spending a larger proportion of their time in the field; instead, they expected that their university degree would help them find an office job in an organization located in an urban centre. The research agenda of the local universities and national research organizations considered mainly technical issues focused on the necessities of the commercial sector. The social research agenda had a tendency to exclusively focus on the theoretical level and did not necessarily apply the generated knowledge to solve practical problems. There were very few researchers or academics who had the time or the desire to spend much time in the field interacting with and learning from the smallholders. This was also true of foreign researchers working in international research or with donor organisations (Pokorny and Johnson, 2008c).

The effect of unjust institutional structures and the globalization process

One of the structural problems hindering the formulation and implementation of alternative proposals for development in the Amazon, is that the strong dynamic linking the global economy with local resource management – including the extraction of resources and the production of raw materials, followed by processing, commercialization and consumption – function independently of local and even national boundaries. Thus, as observed by Stiglitz (2006), we found that the process of globalization impacted all aspects of rural society, albeit to varying degrees depending on the level of development and integration of specific regions into the national and global economies. Consequently, the Amazon and its resources, despite several attempts to protect these resources from exploitation, are generally accessible to economic actors on larger scales, both regionally and globally (Pokorny et al., 2012).

In all the study areas, we found a large number of actors with significant capital, including companies and rich individuals who, often benefiting from incentives for market-oriented policies, used the resources of the region to secure raw materials for their industries, took advantage of the low land prices to purchase areas for cattle ranching or agroindustrial production, utilized the cheap available labour force, or simply turned a profit in land speculation (Pokorny et al., 2012). These actors systematically evaluated and utilized existing opportunities to benefit from the regions' land and resources, thereby creating a strong dynamic that left little room to develop and implement alternative approaches. In such a situation of powerful actors using their means to set the rules of the game in line with their specific interests, it is extremely difficult to implement effective protectionist policies to generate the necessary space for development and experimentation with alternative development models. This is particularly true in the context of the Amazon, where the interests of the powerful elites were historically promoted by unjust institutional structures apparent in the still-widespread phenomena of patronage and political favouritism and the weak and unsteady performance of the public sector (Weigelt, 2011). As described in Chapter 4, before the construction of roads accelerated the land-use dynamic, economic life in the region was organized in a paternalistic system, with a few families controlling both resources and people. In many parts of the ForLive study areas, we found that these systems, in which the families deliver their production at low prices to a patron, intermediary or company who cares for equipment and logistics, persisted.

We also observed strong links between the economic elites and the politicians – who in many cases were united by kinship – in all the case study areas. In Brazil and Bolivia, for example, we observed a strong relationship between local politicians and the companies exploiting timber and Brazil nuts, as well as the large cattle ranchers. Often, the councillors and prefects promoted these actor groups because they were personally involved in commercial activities or because the economic actors were actual or potential donors to their election campaigns. In fact, many local governmental authorities promoted economic policies for their own benefit (Blum, 2009). Similarly, in Peru, we found that companies systematically utilized their political influence to generate and use economic opportunities. For example, the Grupo Romero company, through their house bank Banco de Credito, recently obtained legal titles to several million hectares along the Interoceanic Highway (then under construction) for the production of palm oil

(Elaeis guineensis), peach palm *(Bactris gasipaes)*, and, in the floodplains, sugarcane *(Saccharum officinarum)* (Cáceres, 2005; Defensoría del Pueblo del Perú, 2010).

A ForLive study in Ecuador analyzed these unequal relationships based on favouritism and political connections (Thomas, 2008) and revealed their great importance in political decision making. Unlike the fairly common and well-known corruption at the everyday, local level, this influence was identified as a type of corruption that was manifested in the manipulation of decisions at a strategic level, such as in the formulation of laws and regulations. The study also observed this 'high-level' corruption in the announcement of large projects related to, for example, the exploitation of mineral resources such as petroleum and gas or for the construction of infrastructure. Likewise, in Peru, timber companies massively intervened to adapt the initial forestry concession law (2000) to favour their interests. Specifically, they managed to arrange that various traditional forest practices typically used by smallholders be defined as illegal, thereby favouring the companies' position in accessing the offered timber concessions due to their financial and technological capacity to present management plans consistent with the norms for Reduced Impact Logging. These companies were also able to hinder the creation of a regulatory agency that would have had controlled their concessions, so that their tree inventories were approved without a field inspection. Similarly, in Bolivia, the concessionaires managed to use their influence in the 'Forestry Council', which was a government advisory board composed exclusively of logging company representatives, to pay royalties only for annually exploited areas rather than for the entire concession area, as initially planned. Often, these attempts to influence political strategies were accompanied by discourses favouring the specific interests of powerful actor groups (Medina et al., 2009a).

In the states and municipalities, to guarantee their political advantage and to defend their economic interest, the local elites had to ensure the goodwill of voters among a large proportion of smallholders. Therefore, in addition to the lack of transparency and systematic misinformation, the people in power often propagated favourable reporting in mass media, which were usually also under their control. They also offered jobs, sales and consumption options for local families to win their favour. In the study areas, we found that this scheme operated in several ways, usually through the provision of gifts and individual presents to local leaders, the approval of small investments or loans, or, very commonly, through investments in infrastructure. Many of these favours were used as political campaign tools, especially before elections. Other examples in the study areas included the facilitation of local transport, attention to urgent health needs and support for schools. Each favour was usually framed within a populist discourse. Using these opportunities, a number of new economic actors were able to establish themselves as the new 'patrons'. Often, these powerful newcomers were perceived in a positive light by the smallholders, despite their obvious involvement in corruption and opportunism (Bennati et al., 2008; Medina, 2008; Medina et al., 2009b).

Policy inconsistencies within government sectors

In general, national policies, including those directed towards rural areas, were based upon the realities of the more developed regions within the countries. Thus, the majority of the countries did not have any specific policies for the Amazon that considered the particular conditions of the region. This shortcoming became particularly apparent in comparative studies of the legal frameworks related to the use of forests by smallholders in the Amazon (Sabogal et al., 2008a; Carvalheiro et al., 2008; Ibarra et al., 2008; Martinez Montaño, 2008). These studies revealed a wide range of relevant sector policies with major influences on land-use dynamics, including economic, social and environmental policies. In addition to a number of supplementary regulations that directly addressed relevant issues such as protected areas, water resource management, the rights of traditional communities and indigenous groups and land-use planning, there were also many other sector policies that impacted smallholders' ability to use their natural resources in a more indirect, but often even more significant, way. This type of policies included, for example, regulations related to the decentralization of regional governments, the use of hydrocarbon resources, transport and shipping, customs and foreign trade, administration and civil and criminal rules. The study confirmed that these policies created many conflicts because there was significant overlap – as well as contradictions on one hand, and regulatory gaps on the other hand – with legislation directly related to the use of forests and other natural resources (Pacheco et al., 2008).

In general, the policies that most strongly influenced the land-use dynamics in all the Amazon countries were found in the economic sector and aimed to foster economic growth by serving the needs of capitalized actor groups such as industries and large investors (Pokorny et al., in press b). In fact, almost all the Amazonian countries massively promoted large businesses with lines of credit, subsidies, tax breaks and, more indirectly, the construction of infrastructure, logistical support and international trade agreements. Often, these national policies were financially supported by international cooperation with organizations such as the World Bank, USAID, KfW and a number of private investors. In the region of Santarém, Brazil, for example, the North American companies Cargill, ADM and Bunge were the most important investors in infrastructure for the transport of soybeans. We encountered an egregious example of policies favouring actors involved in global commodity markets in Peru, where the economic inclusion of the Amazon in the production of biofuels was explicitly added to a trade agreement with the US (Tratado de Libre Comercio; TCL). One decree (No. 1015) explicitly designated community areas for the cultivation of agroindustrial products; another decree (No. 1090) promoted the reforestation of degraded areas with species needed for the production of biofuels by large companies. These regulations contributed to the tragic clash between the police and indigenous groups in the city of Bagua in June 2009, in which 22 people died and protesters took hostage and later killed 12 police officers. In response to these events, the Peruvian Congress passed a law giving indigenous people the right to be consulted before any projects are initiated on their land. However, it remains unclear whether this law will confer the right to veto projects, and whether there are sufficient resources for effective implementation.

National policies most directly targeting Amazonian smallholders were generally found in the environmental and agricultural sectors. However, the policies of these two sectors were typically in extreme contradiction because the environmental sector was mainly dedicated to the protection of natural resources, especially forests, and thus the control of their use, whereas the agricultural sector highlighted and promoted production concerns. However, despite being strengthened during the last decades as a priority area for international development aid, the environmental sector still had only marginal importance compared with the other sectors such as the economic sphere (Pacheco et al., 2008). Even the government of the Brazilian state of Acre, marketed as the 'Government of the Forest', which was oriented explicitly towards forest-based smallholder production, still dedicated by far the greater part of its public budget to classical agricultural development, particularly the promotion of cattle ranching (IBGE, 2009).

Social policies targeting poor rural families were generally weak in all these countries. Similar to the marginality of the environmental sector within the overall policy framework, debates and decisions on strategies for fighting poverty only involved a limited group of people who were often distant and isolated from the people and organizations active in the most relevant sectors (Pokorny et al., 2012). The problem of effectively considering smallholders in political decision-making and mainstream policies was further exacerbated by the lack of continuity in governmental organizations. Based on our observations in the study areas, many political initiatives nominally favouring the rural poor were promoted near election time but were not subsequently enacted. Conversely, after each election, particularly when it was won by the opposition, there was nearly always a profound political restructuring process including drastic changes in local political management, especially in the leadership of the various departments that worked directly with smallholders. Consequently, the few policies that were designed to promote smallholders' interests suffered from grave discontinuities and insufficient financing, so they were poorly implemented or simply abandoned.

Weak performance of the state

In all the rural zones of the study regions, governmental organizations – at the federal, state and municipal levels – were present only in rather irregular ways and seldom responded to the demands of the smallholders, often due to a lack of physical presence secondary to limited human and financial resources, generally poor infrastructure, corruption or, as described above, because politicians in the centres of political power at the national and local levels often defended the interests of the economic elites. A dramatic case of the inadequate performance of state agencies came to light in the half-hearted attempts to clarify the issues of land tenure (see Chapter 6). For example, the Land Tenure Institute of the state of Para in Brazil (ITERPA), which was established in 1975 with the initial aim of advancing the legalization of land access at the smallholder level, has since focused mainly on consolidating large properties. An analysis of the land titles issued between 1976 and 2004 showed that only 13% of the titles covered almost 88% of the registered area (ITERPA, 2007). Further evidence that points to the lack of effort in consolidating areas settled by small colonists includes the fact that, in the majority of colonization areas, most basic public services remained unavailable, even decades after the provision of incentives by the state. This problem of perverse political prioritization was aggravated by the notorious lack of human and financial resources in the public sector, which is typical of the countries in this region. Often, the difficult financial situation of governmental agencies was further complicated by the incompetence of the administration in collecting taxes and the lack of implementation and follow-through on the budget, an observation confirmed in the study by Rose-Ackermann (2007).

We found a typical example of this unfortunate combination of lacking political will and insufficient resources in the efforts to control the forestry sector. In reality, despite impressive discourses and the strong engagement of international partners, none of the countries was able to establish effective control mechanisms. In Brazil, at least, the process of deforestation was effectively monitored as a consequence of immense investments in remote control technologies in the field. The majority of control actions, when they existed, were strongly focused on ensuring that users had management plans rather than on identifying those who were engaged in illegal activities. In particular, FSC-certified enterprises complained about the frequency and pettiness of auditing visits by governmental agencies. The use of such methods by the auditing teams was obviously facilitated by the relative ease and comfort of such activities compared with the difficulty of effectively monitoring and stopping companies that were operating illegally. Additionally, with the recent efforts of many countries to decentralize governance structures, this problem has persisted or even worsened. Although there was a gradual transfer of responsibilities from higher hierarchical governmental levels to the state and municipal governments, these processes were generally not accompanied by the necessary investments or restructuring of budgets (Sabogal et al., 2008a; Carvalheiro et al., 2008; Ibarra et al., 2008; Martinez Montaño, 2008).

Another problem that negatively impacted whatever governmental initiative existed to improve the situation at the local level was the low technical ability of the technicians contracted by governmental agencies. These technicians often had not received adequate training, did not keep current with advancing technologies, and did not receive salaries that matched the functions that they performed. Consequently, they often failed to meet their obligations in a responsible way, a problem that was further exacerbated by the lack of resources and highly bureaucratic internal procedures. We found many cases in which the governmental officers responsible for the field work were not able to leave their offices because vehicles or funds for fuel were not available.

Consequently, direct communication between the municipality or district governments and the local families was almost always limited to a few persons related by birth or political ties, thus excluding the large majority of smallholders who did not have the necessary personal connections to engage in contact with the local governments. Their isolation from the public sector was further aggravated by the fact that the majority of smallholders generally had little knowledge about the legal framework, the decision-making process and possibilities for partici-pation, and they did not understand the institutional logic behind the actions of the government (see Chapter 5). It was precisely this combination of lack of contact and knowledge with insufficient consideration and ignorance of local affairs by the politicians that created the generalized perception of an 'absence of the state'. Undoubtedly, the capitalized actors, helped by their relationships with the local politicians, were much more adept at using the space generated by inconsistent and non-transparent regulations and decision processes to realize their objectives while impeding more effective action by the state in favour of smallholders.

High production costs and low prices

The reflection events held within ForLive revealed a broad consensus among politicians, experts and smallholders' representatives that one of the main challenges in guaranteeing an adequate level of well-being for smallholders was the identification of locally viable options for the generation of urgently required income. However, the Amazon region presents difficult conditions for the production of many agricultural and forestry products, resulting in compet-itive disadvantages for Amazonian producers in global markets. Many factors contribute to this situation, most important of which are the high proportion of land with low soil fertility and climatic conditions conducive to pests and diseases. For mechanized agroindustrial production systems and plantation forestry relying on monoculture, these factors result in the need for fertilisers to increase and maintain productivity and pesticides and herbicides to control insects, fungi and weeds. However, perhaps the most important disadvantage of Amazonian producers in commodity markets is the large distance from urban centres and low-quality roads (see Plate 15), which result in enormous logistical challenges and high transportation costs (Pokorny et al., 2012).

Consistent with these hurdles, a study by Hoch et al. (2012) analyzed the financial attractiveness of forestry plantations for smallholders and demonstrated that the lack of attractive markets and high transport costs were the main reasons why only very few smallholders established and maintained tree plantations. Even in the rare cases of families who managed to harvest products from their planta-tions – which they initially established with high expectations – some were not able to sell their products for adequate prices. Thus, the timber prices in the study areas varied from US$2 to US$50 for a whole tree and typically did not justify the large investments in human labour during the establishment and harvesting phases. For many non-timber forest products (NTFPs), the rewards of commer-cialization in distant markets were even less attractive in financial terms and even riskier due to the lack of durability of the products (see Chapter 2). Adequate commercialization was a crucial challenge for literally all the smallholders contacted during the project, including those who used highly diverse agroforestry systems.

With regard to forestry products, the few cases showing stable and favourable marketing conditions were observed mostly in two situations: within ongoing development projects in which NGOs facilitated marketing in niche markets, often located abroad and thus offering attractive premium prices, and for several specific NTFPs that were harvestable annually and could be sold in local, easily accessible markets to generate annual income (Hoch et al., 2012). Like Padoch et al. (2008), we also found that the emergence of rapidly growing urban centres in many regions positively affected the opportunities for smallholders settled in peri-urban areas to market forest and agricultural products. We found strong indications that local markets were generally more attractive to smallholders, particularly those located in peri-urban areas, as commercialization in national or international markets typically depended on the continuous support of NGOs, companies and consumers engaged in social issues (Ortiz, 2007; Shepherd, 2007; Shackleton et al., 2008; Kern, 2012). However, the commercialization of local markets faced enormous logistical challenges and significant power imbalances that enabled the local elites to capture the eventual profits (Kern, 2012).

The challenging conditions hindering competitive production combined with the fact that there are relatively few consumers in the region, with most of them concentrated in the larger urban centres, created a dilemma for small producers. On one hand, these producers urgently needed improved infrastructure to better access markets for the commercialization of their products; on the other hand, after they were connected to markets by roads, producers from outside the region generally benefited from better production conditions and consequently started to occupy the local markets on a large scale, often offering products of better quality with competitive prices (Pokorny et al., 2012). In this situation, it was difficult for the families in the region to produce in a competitive manner. This difficulty became more pronounced as emerging economic opportunities and available subsidies were systematically captured by large producers but not by smallholders. This phenomenon, described by Myrdal (1970) as 'circular cumulative causation', was commonly found in the study areas, for instance in the case of subsidies for ranching and livestock and for the cultivation of soy and palm oil (*Elaeis guineensis*). We found that attractive prices paid on the global market, in combination with massive governmental incentives, contributed to a dramatic increase in the cultivated area dedicated to these crops. However, because of the large investments necessary for machinery, fertilisers, equipment and infrastructure, the smallholders were mostly excluded from these opportunities. In light of this fact, Homma (2006) argues that the only real opportunities with economic potential for smallholders involve production processes that require intensive labour and thus cannot readily be mechanized; examples include coffee, cocoa and, more recently, açaí (*Euterpe oleracea*), palm heart (*Bactris gasipaes*) and patauá (*Oenocarpus bataua*), which yields a high-quality palm oil.

Smallholders' weaknesses

The obstacles to strengthening smallholders' socio-productive systems as a basis for sound local development were not caused only by external conditions. The weakness and vulnerability of the smallholders themselves, grounded in a history of oppression and exploitation perpetuated to this day by unfair institutional structures, interfere with any initiative for locally based development. The majority of the families contacted during the project suffered from insufficient resources, problems with communication and organization, and, perhaps most importantly, the lack of value accorded to their own cultural identities.

Lack of capital

The majority of the families visited in the field had few resources; for these families, investments in urgently needed improvements were almost impossible. This lack of resources significantly limited their flexibility and opportunities for action and innovation (Vos et al., 2009). The families generally used small amounts of liquid capital, when they had any, for the consumption and purchase of small goods and the education of their children, or in emergency cases such as sickness, accidents, or catastrophe affecting their harvest. Often, the natural resources on their land represented the only economic capital that they had, and even this capital was often of limited value. As a consequence of poor settlement planning, many colonists had a large proportion of low-quality soils with little fertility on their land, which in turn led to low productivity. Moreover, missing legal land titles, in combination with the great pressure from external actors interested in land and resources, frequently caused uncertainty about the ownership of resources and the options associated with them, which also reduced smallholders' ability and willingness to make long-term investments (Trejo, 2007).

Reflection on the individual histories of families involved in the ForLive case studies indicates that the modest improvements achieved by the families over time resulted from extremely slow processes and often required great effort (Vos et al., 2009). These families almost always endured repeated crises in their efforts to attain a better life, e.g. fire, floods, droughts, the death of children or the drastic fall of commodity prices, each resulting in serious setbacks or even the loss of any hard-won progress. Many families reported that they felt obligated to sell part of their capital (livestock, a part of their property, etc) that was built through many years of arduous work to overcome crisis situations such as the illness of one of the family members. Consequently, families were forced again and again to start from scratch to improve their living conditions, often with little return. This vulnerability was identified as an extremely limiting factor in their ability to overcome the difficulties inherent in reconfiguring a more just approach to local development. This situation also highlighted a type of chronic dependence on external partners offering opportunities and options for development, even when they sometimes benefited only a small number of families and held illusory promises for the vast majority.

Low degree of communication and organization

Another worrying observation was the low degree of communication and organization of the smallholders, a challenge that has been well analyzed by Selener (1997). The families in the case studies extensively used opportunities to exchange information during common events such as church visits, football matches and house-to-house visits. However, although there were several examples of smallholders effectively organizing themselves, often with external support, the lack of communication evident in the majority of the case studies represents a serious problem (see also Chapter 2). In particular, in the case studies of settlements in which families lived in their individual lots, communication between producers proved to be very difficult, mainly due to the significant distances between houses. Consequently, in these contexts, we found that the exchange of information between families was quite limited, although their heterogeneous cultural background and the profound knowledge accumulated in their places of origin or on the farms where they worked as labourers would likely have offered a great opportunity for local learning about the great diversity of their production practices. The level of social exchange was higher in more traditional contexts. However, in these contexts, we also observed a tendency towards individualism and a surprising lack of communication regarding essential aspects of daily life and production. In several cases, the families did not even know how their neighbours were cultivating their crops. The traditional practice of cooperating with neighbours in small groups for specific tasks or during labour peaks in individually or commonly owned properties (e.g. *mingas* in Spanish and *multirão* in Portuguese) was becoming rare and, in several cases, was even lost.

Contact with persons and organizations outside the smallholders' local network was often very restricted. Indigenous groups in particular showed great resistance to contact with external actors. In general, the isolation of communities due to large distances and difficult access made the exchange of information with external actors scarce. Nevertheless, with increasing access to means of communication, especially the radio, and the expansion of the electrical infrastructure and access to satellite dishes and television, families received increasingly generic information. However, again, this influx of information was directed by outside agendas and interests and strongly shaped by the culture of large and distant urban centres; thus, the content and meaning of these media were controlled by forces in these urban centres (Martín-Barbero, 2006; Steinbrenner, 2011). As a result, a large proportion of smallholder families remained poorly informed and highly susceptible to manipulation. The situation is exacerbated by a lack of educational opportunities, as a direct consequence of the lack of adequate schools, with only a few people demonstrating consciousness of their condition and their rights and opportunities, including the possibility of influencing public policies to their benefit.

Additionally, smallholders' representative organizations were significantly challenged in their attempts to participate in public policy debates (Medina et al., 2009a, 2009b). The grassroots organizations representing Amazonian smallholders in the study areas generally had less national visibility and influence in comparison with representative organizations from other regions. Consequently, the achievements of smallholders' organizations at the national level, e.g. with regard to policies for credit models and technical assistance, were generally better suited to the interests and demands of families located in more developed regions and did not adequately address the situation in the Amazon. Moreover, generally speaking, the national representatives tended to prioritize issues – such as the management of natural resources or norms for economic activities in extractive reserves – that did not necessarily reflect the most immediate needs of Amazonian smallholders. Consequently, although the smallholders in the study areas did have regional representation for their demands, national representation of their interests remains another unsolved challenge.

Lack of valuation of local cultures

One concerning observation made in ForLive was that many families in the study regions did not value their own culture and knowledge. This problem became dramatically evident during the early stage of the project, in the attempt to identify 'promising' experiences of smallholders managing their resources (see Chapter 1). Virtually none of the more than 250 persons we consulted, including politicians, experts and technicians, as well as local farmers and representatives of local organizations, considered any locally developed practices to be 'promising'. Instead, the respondents exclusively described externally promoted projects and programmes, which the locals were intended to manage by adopting the externally promoted technologies. In-depth interviews with the smallholders in the case studies confirmed this lack of valuation of their own practices and capacities. Instead, they tended to place unconditional faith in the 'word of the expert', although they often did not fully understand the explanations of the professionals, who had generally received little training in communicating or working effectively with communities (Montero, 2007).

Although the majority of the families in the case studies expressed strong emotional and physical bonds to their land and lifestyle, they generally were very open to drastic changes in their ways of living and using the land (see Chapter 2). This willingness to change was not only reflected in the high rates of migration and fluctuation out of the community, but also in the fact that the smallholders systematically explored opportunities for greater participating in the 'modern lifestyle'. The families also had high expectations regarding external support, spontaneously accepting the opinions and proposals of the development organizations. Generally, external technicians were perceived as superior and more qualified (Pokorny et al., 2005). The experiments in ForLive designed with the intention of stimulating self-reflection processes and support for local initiatives showed that the expectations and motivations of the majority of the smallholders with respect to collaboration with external agents were dominated by paternalistic and historical relationships. This phenomenon was reflected in the displacement of responsibility for actions, passivity in consequences, a philosophy marked by the receipt of support and a latent belief that only someone from the outside can solve problems (Pokorny and Johnson, 2008c; Flick, 2008).

Obviously, the prevailing development discourses were also ingrained in the smallholders' own minds (Medina, 2008; Medina et al., 2009a, 2009b). In the discussions held during the case studies, we became aware that many families still rely on the historically embedded paternalistic model as their path to development. In this sense, the affirmation of Paulo Freire (1979) that the poor tend to base their imagination of what could be on their experience of oppression was confirmed in many situations, perhaps most dramatically in the veneration of successful politicians known for corruption.

8 Rural development rooted in local culture

Proposals for action

The evidence presented in the previous chapters suggests that smallholders have a significant potential to contribute to sustainable local development in the Amazon. In truth, smallholders do contribute to deforestation but this effect is highly concentrated among individual properties used for small-scale agriculture. Consequently, settled smallholders typically do not affect the large remaining areas of primary forests on common or public property when conditions for colonization are adequate and pressure from commercial actors is absent (see Chapter 3). In areas where local families possess larger plots of land in collectively owned forests and where the governmental authorities manage to effectively protect public forests, there may exist the potential for smallholders to create environmentally stable landscapes that generate economic opportunities for local development (see Chapter 3). Currently, however, smallholders are undergoing a process of cultural homogenization that originates in a vision of development oriented towards the commercial exploitation of resources dominated by capital-endowed actors and based on a legal-institutional reality that reflects historically unjust structures (see Chapter 4).

This dynamic is characterized by a gradual accumulation of production factors in the hands of external actors with interests in the regions' natural resources including forests, minerals, energy and, most importantly, land for agricultural production (see Chapter 4). Taking advantage of their capacities and power, these influential actors, who are often linked to attractive export markets, occupy the land and resources or manage to oblige smallholders to use their land to produce for global production chains. For many poor families, participating in this process represents one of the few options for improving their precarious situation, and so they perceive the construction of roads, involvement with commercial actors and the orientation towards emerging commodity markets as perhaps the only opportunity to generate income, which is generally understood as a necessary precondition for obtaining access to public services such as education and health and participating in consumerism. This approach, however, implicitly demands that the smallholders, as in past colonial times, increasingly replace their cultures and traditional ways of using natural resources with models of production and organization developed and induced by external actors and institutions. These models, with rare exceptions, follow global demands and concerns and generally focus on a limited number of internationally traded commodities, with systems of production that require significant investments; they are oriented towards a large economy of scale and require the establishment of hierarchically organized business systems highlighting operational aspects to guarantee productivity and profit. Additionally, the efforts of national governments and international communities for the demarcation of protected areas are generally based on primarily external demands and initiatives.

Consequently, we observed an ongoing process of deterioration and marginalization of the cultural diversity of Amazonian populations that indirectly contributes to the process of extensive deforestation and environmental degradation. This situation again puts more pressure on local families, exacerbating the process of migration and urbanization, and all the consequences of social disintegration and degradation that are associated with these processes (see Chapter 4). Although they generated crucial benefits for many families, the numerous initiatives initially aiming to create just and sustainable local development (see Chapter 5) tended to accelerate this process of smallholder marginalization. This marginalization occurs because external actors invest exclusively in the adaptation of families and communities to a model of development cast in the logic of the market and conservation (see Chapter 6).

It has been argued that a certain degree of cultural deterioration and environmental degradation must be accepted as a necessary cost of economic development, which generates urgently needed benefits, especially for the poor rural families themselves. However, when the manifold social and environmental consequences – namely, the loss of cultural diversity and traditional knowledge, the degradation of nature, social unrest and the consolidation of an unjust social order – that are associated with this shift continue, it seems useful to consider alternatives for the design of development policies that may be more compatible with the social, cultural and environmental diversity of the region.

One of the main lessons learnt from ForLive is that the culture, capacities and demands of the huge range of indigenous groups, traditional communities and small-scale colonists in the region can serve as a reference point for constructing an alternative, fairer and more sustainable model for local development. This model inherently involves a shift in starting points, from seeing these people as the 'problem of development' to seeing them as having the potential to achieve alternative development. However, to realistically assess this fascinating possibility, it is important to be conscious of the various barriers that impact the viability of this innovative proposal and place smallholders at the centre of focus (see Chapter 7). Such a development approach implies drastically rethinking several of the prevailing conceptual paradigms and discourses and carefully exploring and evaluating the possibilities for implementing such a vision in practice.

Considering that efforts for rural development in the Amazon during the last few decades have been almost exclusively based on the classical development paradigm of economic growth catalyzed by market dynamics and accompanied by command and control measures to ensure social and environmental safeguards, attempts at formulating and seriously assessing ideas, strategies and tools for the realization of an alternative model of development are only in the beginning stages. However, using the findings from ForLive as a foundation and referring to the vast available research on endogenous development (Albuquerque, 2001; Vásquez-Barquero and Garofoli, 1995; Vásquez-Barquero, 2002), political ecology (Bryant, 1998; Blaikie, 1999; Neumann, 2008; Nygren and Rikoon, 2008; Rocheleau, 2008) and the chronicles of power (Green and Hulme, 2005; Harriss, 2007), in this section, we suggest three basic proposals for action: (1) instead of undertaking projects *for* smallholders, smallholders should be supported in undertaking their own projects; (2) locally existing capacities and cultures should be strengthened and valued; and (3) instead of adapting smallholders to externally determined development models, efforts should be made to adapt the legal-institutional context to facilitate socio-productive models locally developed by the smallholders.

Instead of undertaking projects for smallholders, smallholders should be supported in undertaking their own projects

In the Amazon region, development programmes and projects targeting local people are generally designed by external actors, including governments, donors, NGOs and scientists. These external actors define the objectives, make decisions about resources and manage the logistics of the project. Consequently, the locals typically come into the picture after the programmes and projects have already been started. Even in initiatives explicitly specifying more participatory approaches and seeking to establish a dialogue between external and local actors, the locals are commonly perceived as merely recipients of external support, including the families themselves. This perception contributes to serious shortcomings, often confirmed in the observed development initiatives, such as the definition of objectives without consideration for local priorities, inadequate project duration periods, requirements exceeding local capacities and a lack of local control over the application of resources, among others.

In view of this problem, it may be promising to forsake the classical diffusionist approach of technology transfer and capacity building. This would require departing from the emphasis that asks how to perform projects for smallholders and instead emphasizes exploring ways to help the local families in conducting their own projects within which they can define their own priorities, consider their capacities and ensure that the resources are used in accordance with their interests and needs (Pokorny et al., 2005). As described by Chambers (1987), when such an approach is taken, the families stop being simple recipients of external support and instead become responsible managers of their own development. Following the teachings of the great Brazilian educator Paulo Freire (1979), the local families should become the protagonists of their own development and, correspondingly, this development must be grounded in awakened local capacities for defining, managing and organizing endogenous development processes in line with the locally relevant social, economic and environmental features.

Without doubt, such an approach of local leadership in the institutional partnerships with development organizations requires a significant change in the way that development agents and the smallholders themselves interact. In particular, it involves two operational elements: first, the design and establishment of mechanisms to mobilize the local families to think about ways to optimize their own socio-productive systems – implying, in many situations, a facilitated process of awareness-building and political consciousness; second, a dramatic shift in how the external support system works, shifting from a system seeking to apply many resources during a short time period to a system of more extensive but continuous accompaniment of the local families with a focus on applying human and financial resources in a way that is compatible with local realities (Pokorny et al., 2005; Pokorny and Johnson, 2008a, 2008b). An excellent example of this approach is the 'Campesino-to-Campesino Programme' described by Selener (1997), which was introduced in Central America in response to the failure of conventional top-down approaches for technical assistance and rural extension that were promoted by the public authorities. In the farmer-to-farmer approach, the families start to reflect on their reality to construct a development process that has a starting point in their own experiences and capacities, with the ultimate aim of promoting the dissemination of locally developed technologies that work and are suitable for their specific realities. The approach instigates the formation of 'promotional' volunteers who are capable of kicking off an agricultural trans-formation process driven by exchange and communication among families. The role of the external professional is mainly that of creating adequate conditions for exchange and reflection. From the beginning, the facilitator encourages locals to assume leadership roles in the process and assume responsibility for developing the capacities needed to move the process forward. Thus, one of the principal tasks of the external facilitator is to identify local 'promoters' to replace him or her in the facilitator role.

Additionally, initiatives focusing on the establishment of mechanisms for local governance depend on the local capacity of residents to effec-tively organize their socio-productive systems on their own (see Chapter 5). Basically, in this approach, the smallholders are expected to establish, through common consensus, the rules and mechanisms for the regulation of access to and use of their resources to guarantee effective and long-term function. Distilling the work of Ostrom and her colleagues (Ostrom, 1999; Gibson et al., 2000) in light of the insights gained in ForLive suggests that this process may proceed through three operational phases: (1) the identification of existing local agreements based on common interests regarding access to and use of resources and/or the promotion of a process for the creation of such agreements where they do not exist; (2) the establishment of locally organized mechanisms and authorities that allow families to discuss and formally agree on the rules and procedures and through which their implementation and fulfilment can effectively be controlled; (3) the legal recognition of the above rules, procedures and instances by the governmental authorities, which can then use their power to support the local mechanisms by enforcing the regulations and protecting them from external disturbances. Historic examples and more recent experiences with this type of local governance (Aghón et al., 2001; Altenburg and Meyer-Stamer, 1999; Pokorny et al., 2005; Scott and Garofoli, 2007) show that the consideration of the interests and participation of all relevant actor groups in the elaboration of these local governance systems not only has the potential to stimulate the effective organization of resource use in collective areas, but may also promote the management of individually owned areas in a way that is acceptable for all. However, such an approach requires not only the devolution of decision-making power, but also the transfer of sufficient funding from the upper governmental levels to local agencies. This approach also requires the support of local agencies in enforcing their rules against uncooperative individuals and groups, especially wealthy and powerful actors.

Strengthening locally existing capacities and the valuation of local culture

The results of ForLive confirm the observations of studies undertaken in other contexts addressing the great potential of local livelihood strategies and diversified small-scale production systems to achieve good and stable living conditions for these families and provide a basis for the sound social, economic and environmental development of rural regions (Ellis, 2000; World Bank, 2007; Wiggins et al., 2010). This perspective views smallholders as being autonomous, possessing knowledge and exerting control over technology, making decisions and managing production and social processes. Thus, locals are not simply adopting the business model of modernized agriculture that is oriented towards integration into global commodity chains, but rather they consciously choose their own way of living and working in the framework of a capitalist society (Schneider, 2003).

The insights gained during the project indicate that one of the main difficulties in achieving such locally based development is that the smallholders themselves tend to ignore the value and potential of their cultures and knowledge (see Chapter 7). One reason for this situation was found in the colonial history that has determined and often continues to determine hierarchical and paternalistic relationships and roles in this region (see Chapter 4). However, in modern times, this logic is perpetuated in the discourses of progress and rural development, which implicitly assume the inferiority of the smallholders in relation to other actor groups, particularly professional companies and capital-endowed actors with the capacity to fulfil the requirements of global capital and commodity markets; moreover, the logic of international agendas for environmental protection and poverty alleviation are still dominated by industrialized countries (Pokorny et al., in press b).

In this situation, it seems essential to build awareness among the smallholders that not only is their culture valuable, but that their own knowledge and capacities can be the basis for economically effective and environmentally sound development. This perspective is especially justified given that most smallholder families, in contrast to people living in industrialized urban centres, practice an acceptable level of consumption considering the world's limited biocapacity (see http://www.footprintnetwork.org/en/index.php/GFN/). In view of Amazonian smallholders' ecological footprint, their production schemes may serve as a better reference for sustainable development than the normally referenced lifestyle of 'developed' countries. This goal would require efforts aimed at raising consciousness among local communities regarding their roles as more than socially isolated groups who are subject to decisions made by powerful external actors. Instead, these communities should be active participants in civil society, with the possibility and the capacity to influence public policies in accordance with their particular interests and demands. The importance of this process of local mobilization and political consciousness is reflected in the fact that, in the case studies, initiatives that were more strongly influenced by smallholders showed better economic and social performance than initiatives in study areas dominated by capital-endowed actors (see Chapter 3). To strengthen the political and societal positions of smallholders, it is necessary to systematically invest in two areas for action: first, the establishment of a system for rural education and political awareness-building based on curricula developed in a participatory manner by the local actors themselves (Arroyo et al., 2004); and second, the implementation of effective mechanisms for communication and participation, allowing the smallholders to more directly exchange and interact with the decision makers and thus contribute more actively to the formulation of public policies (Green, 2008).

Educational systems focused on the realities and demands of smallholders

The formal education systems found in the four study areas teach curricula that were developed in urban centres for an urban audience. Consequently, the contents transmitted to rural schools have little to do with rural reality, and the curricula indirectly devalue the rural way of life and the cultures of the smallholder families. In addition, the rural school system, with its fixed time schedules and dependence on the regular daily presence of the students, suffers from significant logistical incompatibilities with smallholders' local realities, thus limiting their accessibility (Hage, 2004). Consequently, there is a high degree of illiteracy and abandonment of school; at the other extreme, many talented young people abandon their families and the rural context itself to attend schools located in the urban centres (UNDP, 2010).

The Casa Familia Rural (CFR; Agricultural Family Schools) in Brazil constitutes an example of an alternative model of education (CEETEPS, 2000). CFR is an educational project that is based on developing curricula relevant to the capacities, necessities and opportunities present in rural contexts while also meeting the demands of national curriculum guidelines. CFR follows the pedagogic philosophy of 'alternating', which intends to better respond to the needs and problems of families living in the rural Amazon. This system is called 'alternating' because it allows students to attend part of the supervised study time in school but also spend much of their 'school' time at their families' homes, where they apply and share their education with their family and neighbours. This learning process is organized in cycles, wherein the students conduct research on local contexts and problems and share and discuss their findings, both with their colleagues in school and at home with their families. Specific worksheets integrate the two types of curricula. CFR had its origins in France in the period between the two World Wars and was introduced 40 years ago in southern Brazil. Since then, more than 25 CFR institutions have been established in the Brazilian Amazon, the majority of which are concentrated in the region along the Transamazon Highway in the state of Pará (Pokorny and Johnson 2008b). We found a number of other similar initiatives in the project region that had elements in common with CFR, such as the Centros de Investigacion Agricola Local (Centres of Local Agricultural Investigation, developed by the Centre for International Tropical Agriculture; CIAT) in Columbia and also promoted in Bolivia, Peru and Ecuador. These centres offer a framework that strengthens the experimentation and learning capacities of smallholders to permit the development of their organizational and management skills. Another example in the agricultural sector are the Escuelas de Campo (Country Schools); this project was originally developed in Asia to promote the comprehension of the agro-ecological aspects of the rice production cycle and was later adjusted and promoted by the International Potato Centre (CIP) for potato cultivation in the Andean region. The Escuelas de Campo promote smallholders' capacities to observe, question and formulate research projects related to their production systems.

Although this type of educational approach has successfully proven its enormous potential in numerous experiments over several decades, especially in the Andean region (Gallagher, 1999; Milani Filho et al., 2002; Texeira et al., 2008), it still plays only a marginal role in the public education systems present in the rural Amazon. Obviously, there are significant geographical and institutional barriers limiting further expansion of these experiments and hindering a long-overdue reorganization of the rural education system (Hage, 2004). In particular, the authorities of the environmental and forestry sectors contacted in the study areas were still far from making use of this immense potential for fostering sustainable use of forests and natural resources for the benefit of the local population.

Promotion of communication

Another crucial component of the mobilization of smallholders' potential for contributing to sound local development is communication (Freire, 1983). As described previously, direct communication and exchange among smallholders is of crucial importance for the diffusion of experiences, reflection, the optimization of practices, the development of new ideas and group organization to defend their interests against other actor groups (Selener, 1997). Thus, the opportunity for smallholders to organize meetings and to visit other people and organizations of interest is of utmost importance in accomplishing any sort of endogenous development.

Smallholders also need access to new means of communication, specifically cell phones and the internet. Access to these new technologies is fundamental, not only in cases of emergency, for example, in the case of accidents, but also to organize collective action and as a source of information to support decision making, for example, in determining the price of a product or the feasibility of its delivery. Platforms or social networks such as MSN, Orkut, Twitter and Facebook would allow smallholders to be more linked into the exchange of information and experiences of people with similar interests and affinities. Simultaneously, such access provides smallholders with the opportunity to increase their volume of knowledge and supports their social integration. Having access to email services, then, is a key requirement for establishing communication channels between subjects of similar levels of power as well as with people of different levels of power in different spheres. However, despite the fact that the majority of these technologies are already widely distributed and the existence of several initiatives designed to link smallholders to these sources of information and communication, the gaps between urban centres and rural zones are immense and increasing (Tauk Santos, 1998, 2000).

Finally, participation in mass communication plays an important role not only in ensuring free access to information propagated via newspapers, magazines, television and radio, but also in securing communication as a human right; people should have the power to influence or actively participate in defining what is being transmitted. Thus, communication should assume a strategic role in the design of rural development. In fact, communication has been widely noted as a social mediation tool in the emblematic case of community radio, which was a growing phenomenon not only in the study areas, but also in many places where there are material or institutional shortages (Steinbrenner, 2011). There are many cases in which the community radio station enters the scene and positions the debate with a focus on the 'slow formation of new public spheres [and] new forms of social imagination and creativity' (Martín-Barbero, 2006). In the study areas, generally composed of municipalities and districts with populations of less than 30,000 habitants, the local representative organizations had often become the main operators of these stations, putting on their own programmes to disseminate their points of view and information about relevant themes. In fact, these community radio stations were often the only communication mode in rural localities, so they generated a strong and large influence as much for the founders of the radio station as for the community audience. Sometimes, radio has been used as a strategic tool in local power struggles for groups that normally lack a voice to gain an opportunity to respond to more commercial forms of mass media, in particular television, that are under the control of the political and economic elites (Steinbrenner, 2011). Where these forms of mass media strongly dominate the construction of 'common sense' in society, Amazonian smallholders still do not have access to a voice, and thus depend on the very few journalists interested in smallholders' views.

Instead of adapting the smallholders to externally defined development models, adapt the context to the socio-productive models of the smallholders

Since the 1980s, there has been an increasing consciousness about the role of territory for rural development. In this understanding of local development, territories become agents of social transformation and not merely a physical resource that supports and provides objects, activities and economic processes (Vázquez-Barquero, 2002, 2010). This viewpoint implicitly recognizes the necessity of integrating marginalized or excluded populations in the general process of national and regional development. In contrast to the current policies principally aimed at the simple assimilation of these regions into the national economies, experiences from rural areas in industrialised countries highlight the importance of valuing the natural and economic resources and the social and cultural capacities available in a determined territory to ensure local benefits in this process (Scott and Garofoli, 2007; Vázquez-Barquero, 1993; Stöhr, 1990).

Another important shortcoming of the conventional strategy for local development that aims to strengthen local capacities is that this perspective implicitly passes the burden of change to local families. In light of the costs of development initiatives, this focus causes immense transaction costs for the families who are expected to adjust and change their socio-productive systems in accordance with development organizations' proposals, a burden that is often overlooked. Investments and innovations are expected from the actors who are least prepared to address these challenges. As a consequence, most of the families, especially the poorest, have little chance to overcome their competitive disadvantages vis-à-vis better endowed actor groups. Thus, it seems worthwhile to consider shifting a higher proportion of the burden of change to actors with good qualifications, sufficient capital and less-vulnerable positions (Pokorny et al., in press a).

This view suggests that, rather than continuing to attempt to adapt smallholders to a development model defined by actor groups and experts located outside the region, it may be more promising to think systematically about the possibility of adapting the legal and institutional context to the reality and capacities of the local people. This adaptation process must go beyond considering simply how to change and modernize local socio-productive systems. In fact, it is crucial to search for ways to strengthen smallholders within their local reality to enable them to continue to engage in their lifestyle and resource use, but with a quality of life that is acceptable for the families, especially the children and the young.

Moreover, in this vision, to avoid continuing down the same path of cultural and environmental destruction that is occurring in the Amazon, it is necessary to provide priority to the smallholders over the large-scale agribusiness companies (Pokorny et al., in press b). This requires protection against the cultural and environmental homogenization that characterizes the influence of the private sector. In addition to the application of strong fiscal policies against the aggregation of land tenure and the effective enforcement of environmental and social regulations, this strategy for endogenous development would also require significant investments in the consolidation of rural areas, including the establishment of public services of adequate quality, the formalization of rights for local populations and the implementation of agrarian reform where necessary. This approach to development also requires acceptance of the degree of environmental transformation caused by smallholders within their individual properties and community areas as an implicit part of their small-scale socio-productive systems (see Chapter 3). Considering the competitive disadvantages of smallholders in the commercial production of agricultural and forest products for international markets (Pokorny et al., 2012), it is also necessary to consider the smallholders as providers of social and environmental services and not simply as members of global production chains. Therefore, it is important to evaluate the possibility of an economically differentiated treatment of smallholders that guarantees fair prices for local products, addresses the establishment of production chains organized in line with diversified, low-input production systems and other innovative ideas such as the payment of basic incomes, municipal protectionism or property taxes (see Box 8.1).

As outlined by Teubal (2009), it is time to invest more significantly in a systematic search for development alternatives that emphasize the autonomy of smallholders, local governance systems and respect for traditional cultures and the environment. In contrast to many other regions in the world, the environmental and cultural richness of the Amazon and the relatively low demographic pressure combined with the size and diversity of the ecosystems that remain largely intact present a unique opportunity for identifying forms of development that are more compatible with the specific situations of the smallholders. It is necessary to carefully explore possibilities for establishing legal and institutional frameworks that better meet local capacities and interests, considering questions such as: How can the management and commercialization of natural resources be better controlled by smallholders? What commercialization schemes can be promoted locally? How can public services be improved to better satisfy the demands of the smallholders? How can the participation of smallholders in local politics be increased? How is it possible to control the powerful actors more efficiently? Given the previous modest efforts by experts, scientists, politicians and smallholders to evaluate these issues, a more intensive debate on these questions has great potential to generate significant new insights, which would help to define development strategies for the Amazon region more appropriately than the current approaches, which are strongly based in the logic of economic growth and driven by the forces of global markets, financial power and attempts at command and control.

Box 8.1 *Valuation or coercion: A conceptual reflection on Payments for Environmental Services (PES)*

Paradoxically, the lack of valuation of the smallholders' potential for a sound local development approach is also apparent in those initiatives that explicitly target the utilization of this potential. One pointed example was observed in initiatives that fall under the rubric of Reducing Emissions by Deforestation and forest Degradation (REDD+), in which international agencies, acting through national governments, intend to devote considerable efforts and financing towards establishing mechanisms for the conservation of forests through Payments for Environmental Services (PES) to landowners and forest managers. In accordance with the general approach of linking environmental goals with a social agenda, a portion of these investments is explicitly dedicated to activities with smallholders.

However, in line with previous initiatives for environmentally sound local development, instead of giving value to the smallholders' existing socio-productive systems as a viable option that could serve as a barrier to deforestation on a large scale, the new payment schemes also categorically seek ways to change and control the behaviour of the smallholders. Not only does this strategy fail to value the functional components of the smallholders' existing systems, it also introduces costly and time-consuming procedures for regulation and monitoring and coerces the application of new technological and organizational systems that often do not fit smallholders' realities (Sato, 2010; Masias, 2011; Pokorny et al., in press a). This situation results in elevated transactions costs for the smallholders, increases external dependency and can seriously affect locals' autonomy and flexibility. In this context, many indigenous and traditional groups have expressed strong resentment against REDD+ and related initiatives (TEBTEBBA, 2008).

Alternatively, PES schemes could be grounded in the respect and evaluation of smallholders' contributions to environmental and social stability. Such an approach would profoundly change perspectives. Instead of paying the smallholders to leave their forests untouched or to manage forests in accordance with externally defined regulations, the families would be compensated for their low-impact productive activities and their strategies for environmental consolidation of their properties that they are already undertaking in a challenging context, accepting significant shortcomings in quality of life compared with families living in urban areas. Such an approach could make funds available that do not require the generation and exchange of carbon credits and the application of strict 'one-size-fits-all' methodologies and would thus allow smallholders to benefit much more effectively from the programmes without requiring significant lifestyle changes or provoking the impression of welfare payments.

Final considerations

The two general approaches to rural development that are gaining ground in the Amazon – at one extreme, development driven by large enterprises and capital-endowed largeholders, and at the other extreme, the conservation of resources by the state in a sort of command-and-control governance – have both proven incapable of ensuring sustainable development for the benefit of the local populations. The studies in ForLive confirm that these approaches – also reflected in the numerous development initiatives targeting smallholder families and communities in the rural Amazon through the promotion of 'Community Forestry', tree plantations and agroforestry systems – have improved the precarious situation of many of these families, but also contribute to the cultural and environmental homogenization of the region, thereby marginalizing the smallholders' approaches to natural resource management and social organization.

Currently, there is a broad consensus that the meaningful integration of smallholders in efforts to develop the rural Amazon is necessary to halt the current destructive land-use dynamic and achieve equitable and sustainable development. However, governments continue to focus on policies that are expected to contribute the most to economic growth and thus still favour capitalized investors as the engine of rural development. This imbalance is exacerbated by a corporate sector that uses its own capital and political influence to shape regional development trajectories. Consequently, current efforts at integrating smallholders rely merely on engaging them in global commodity markets. Obviously, an apparent choice is being made in favour of economies of scale. In practice, the ultimate potential of smallholders to contribute to the sustainable development of the Amazonian is rarely exploited.

There is no doubt that the Amazonian smallholders' practices also cause severe environmental effects and that their production systems leave much room for optimization. The diversity of smallholders' systems must be taken into account along with their differential environmental and social impacts. In this regard, it is important to note that the research presented in this book focused on so-called 'promising' smallholder initiatives. Thus, this study never intended to comprehensively describe and assess the wide range of smallholder realities in the Amazon but, in contrast, concentrated on carefully selected case studies that promised to generate a better understanding of the potential of the various smallholder groups and support policies. Therefore, it is beyond question that the vast majority of families and communities in the Amazon are confronted with worse conditions, fewer benefits from support initiatives and greater threats from the frontier politics of capitalist actors and globalization, and that, partly as a consequence of these factors, they employ socio-productive systems with higher technical and organizational deficiencies and more unfavourable environmental and social effects.

However, while taking this larger picture into account, we found clear evidence demonstrating the great potential of smallholders to contribute to sustainable local development. Our findings suggest that the economic, social and environmental trade-offs in the smallholders' socio-productive systems may, under suitable conditions, be more favourable than those of the majority of the other commercial actors currently dominating the land-use dynamic in the region. In view of this potential, it seems worthwhile that the search for strategies for a more sustainable development of the Amazon should consider the cultural diversity of colonists, traditional communities and indigenous groups as a point of reference. Understanding smallholders as a part of the solution rather than as an obstacle or a problem in sound rural development would provide new opportunities for action.

However, such an approach is challenging. Maintaining the diversity and enormous cultural and environmental potential of the region requires rethinking current visions, concepts and operational modes for rural development and implies the need for drastic changes to the prevailing development paradigms. Instead of continuing to require families to adapt to the demands of the classical model of development, it must be determined how to adapt the legal-institutional model of development to the demands and capacities of the smallholders. The responsible agencies have an important role to play in this process. Smallholders need the formal recognition of rights to their land and resources, which can be facilitated by second-order organizations representing their legal interests in public policy decisions. Infrastructure investments could target smallholders by expanding less expensive roads that reach more communities or by increasing opportunities for transportation by water. Investment in infrastructure is required to facilitate the marketing, storage and processing of the diverse basket of agricultural and forestry products typically produced by locals. Administrative steps that are required to obtain loans, recognize property and sell agricultural and forest products must be simplified, and education and information transmission must be adjusted in a similar manner.

To reach this ideal, more coherent and aggressive policies are needed to protect and reward the socio-environmental activities of smallholders and consequently constrain the influence of the powerful actors with strong commercial interests in the land and resources of the Amazon. Even before these more concrete steps, a drastic change in thinking is necessary. Thus, policy makers, donors, experts, researchers and technicians at all levels must adopt a new attitude of respect and value for these families who have managed to develop and run surprisingly effective social and productive systems, often under extremely difficult conditions.

In the course of the debate on climate change, and with the crises of the global financial system, increasing numbers of interesting development proposals are emerging that consciously rely on the idea of strengthening rather than replacing smallholders' capacities to effectively manage their resources and social networks. However, despite a general but vague awareness of the huge socioeconomic and environmental potential of this region and the proliferation of related discourse, such initiatives still remain exceptions, and governments and research and development organizations widely fail to provide the necessary support. The unsatisfactory consideration to date of development models based on the interests and capacities of smallholders in policy, science and practice, however, brings hope that an intensification of thinking and discussion in this vein has the potential to deliver new and valuable insights, ideas and stimuli.

Considering the serious implications of such a smallholder-oriented development approach for the powerful actor groups that are disproportionately benefiting from the current situation, massive opposition is to be expected.

Naturally, any attempt to influence the current development dynamic must overcome the paternalistic structures of rural society and political favouritism that are prevalent within the current governance systems in the majority of the Amazonian countries and in many international processes and mechanisms. In fact, this is an enormous challenge, most clearly apparent in the endless fights of social movements to gain land tenure and human rights in the area. Thus, it is perhaps more realistic to begin by better understanding smallholders themselves and promote a strategy with a clearer and more critical vision of change, where everyone – researchers, politicians, development agents, entrepreneurs and smallholders – continually explores the possibilities for individual action within personal limitations. There is a great deal of pressure and very little time.

Appendix: Individual research projects linked to ForLive

Researcher	Title	Institution
PhD		
Benneker, Charlotte	Influence of institutional environment on the development of community forest enterprises in Bolivia	University of Wageningen
De Koning, Jessica	Reshaping Institutions. Bricolage Processes in Smallholder Forestry in the Amazon	University of Wageningen
Godar, Javier	The environmental and human dimensions of frontier expansion in the Transamazon Highway	University of León
Hoch, Lisa	Do smallholders in the Amazon benefit from tree growing?	University Freiburg
Medina, Gabriel	Moving from dependency to autonomy: an opportunity for local communities in the Amazon frontier to benefit from the use of their forests	University Freiburg
Robles, Marco	Ecosystem Goods and Services from land use systems of smallholders in the Amazon	University Freiburg
Weigelt, Jes	Reforming Development Trajectories? Institutional Change of Forest Tenure in the Brazilian Amazon	Humboldt University Berlin
MSc, Diploma		
Andersen, Lars	The influence of land use on the soil fertility in the Ecuadorian Amazon	University Freiburg
Billard, Cécile	The influence of social and economic parameters on landscape dynamics and land use changes in the Amazon at the household-level using remotely-sensed imagery. A case study in the community of Palmares II, Brazil.	University Freiburg
Carrera, Robson	The impact of forest management on the floral composition of the remaining standing forest	Universidade Federal Rural da Amazônia
Campos, Paulina	Societal change in Amazonian communities as a consequence of the intervention of development organizations: results from the study of the indigenous community of Callería in Peru	University Freiburg
Chaves, Martha	Indigenous Communal Forest Management: A Study of Influence and Decision-makers	University of Wageningen

▶

Researcher	Title	Institution
De Bruin, Brechtje	The influence of non-governmental organizations on the business strategies of community forest enterprises in Bolivia	University of Wageningen
Del Solar, Reguera	*Discourses on certification of forest enterprises:* Bolivian lowlands	University of Wageningen
Depzinski, Tina	Criteria of rural development organizations for selecting local project partners: A study in the Peruvian Amazon	University of Goettingen
Flick, Kate	Clarifying the understanding of learning and participation: an exploratory survey of participatory forestry approaches in the Amazon	University Freiburg
Kern, Maren	Uncovering local markets for Amazonian smallholders: The case of the Extractive Reserve "Verde para Sempre" in Porto de Moz in Brazil	University Freiburg
Kuiper, Doenja	Small decisions – big consequences; decision-making regarding forest use by smallholders	University of Wageningen
Pantoja Martins, Deryck	New roads and old practices: Community-company Forest management agreements, the case of the Machadinho Extractive Reserve Rio Preto-Jacundá Machadinho d'Oeste – Rondonia, Brazil	Federal University of Pará
Masias, Katia	Challenges of company-community partnerships within REDD+: A case study of Brazil nut concessions in Madre de Dios, Peru	University Freiburg
Montero, Juan Carlos	From knowledge transfer to knowledge exchange: Analysis of smallholders' and professionals' perceptions on tree growing in the Amazon	University Freiburg
Nalvarte, Jaime	Impacto del manejo forestal con fines maderables aplicado en la Comunidad Nativa Callería, Región Ucayali-Perú	Universidad Nacional Agraria La Molina
Romero, V. Diego	The challenge of implementing good ideas: a case of study from the forest sector in Bolívia	University Freiburg
Schlicke, Cindy	A socio-linguistic study of the smallholders in the Brazilian Amazon with special reference to the language pair Brazilian Portuguese – English	University Leipzig
Sonia Ortiz	Potenzial von Märkten für Waldprodukte von Kleinbauern: Ein Fallbeispiel aus Riberalta, Bolivien.	University Freiburg
Sools, Rik	Local perceptions on land degradation in Bolivia	University of Wageningen
Thomas, Jodi	REDD governance: corruption as a catalyst for deforestation in Ecuador.	University Freiburg
Trejo, Karol	Critical events: What are the reasons for change in natural resource management in the Amazon?	University Freiburg
Trümmer, Anne	Expectations of small farmers in the province of Morona Santiago, Ecuador, on use of forests for income generation to improve livelihoods.	SFA
Villacis, Elena	Why producers do not participate in organic certified agriculture? The case of small farmers in the Amazon Highway, Brazil.	University Freiburg
Van Ham, Chantal	Opinions on the implementation of the forest legislation by smallholders in the Ecuadorian Amazon	University of Wageningen

Researcher	Title	Institution
van Velde, Wouter	The interface between the new forest law and Chiquitano life-world: community-based commercial forestry in Bolivia	University of Wageningen
BSc, Specialization		
Aguilera Jiménez, Ana Ruth	Biological and economic evaluation of traditional harvesting practices of sangre de grado (Croton draconoides) in the northern Amazon of Bolivia	University Beni
Blum, Benjamin	Improved diffusion of innovations through locally controlled development projects – The case of soil conservation in Medicilândia at the Transamazon Highway	Hogeschool van Hall Larenstein
Davila, Erick	Evaluation of the impact of institutional policies related to forest management on community development in the indigenous community of Callería, Ucayali region, Peru	Universidad Nacional de Ucuyali
del Aguila, Erick	The economic contribution of agroforestry systems with camu camu (Myrciaria dubia) in a family farm in Yarinacocha, Peru	Universidad Nacional de Ucuyali
Dorado, Marco Antonio	Firewood consumption in the industrial and urban sectors from rural communities in the Riberalta municipality, Beni, Bolivia	University Beni
Escalera Muchía, Edgar	The effect of tree density on cost effectiveness of systematic commercial forest inventories of tropical forest in northern Bolivia	University Beni
Ferreyros Sánchez, Juan Pablo	Evaluation of the factors that influence the land use patterns based on the proximity to the town of Campo Verde, Ucayali, Peru	Universidad Nacional de Ucuyali
Quispe, Claribel	The relationship of women and forest management in northern Amazon region of Bolivia	University Beni
Lange, Nina	Contribution of Medicinal Plants to the Primary Health Care of Indigenous Communities; Case Study of Two Shipibo-Conibo Communities in the Ucayali Region, Peru.	University Freiburg
Maerz, Dorothea	Effects of road construction on a long-term agro-forestry initiative in the region of Puerto Maldonado in the Peruvian Amazon.	University Freiburg
Masaquiza Garzon, Rosa	Analysis and characterization of the exploitation of palm fibre, (Aphandra natalia), in Chimimbimi, Ecuador	SFA
Merino, Juan Pablo	Socio-economic comparison of two different forms of forest use of Pigüe (Pollalesta sp.), Ecuador	SFA
Peralta, Carmelo	The effect of commercial exploitation of Majo (Oenocarpus bataua C. Martius) on the structure and population density of this palm in three communities in the Vaca Diez province of northern Bolivia.	University Beni
Soares, Luciane	Production cycles in the Amazon: Limits of agroextractivism for smallholders in the production of asai (Euterpe oleracea), in lower Tocantins region of Pará, Brazil	Federal University of Pará

Researcher	Title	Institution
Tonore Freitas, Carlos Alberto	Study of the production and marketing of the fruit of Majo palm (Oenocarpus bataua C. Martius) in two small farmer communities, El Hondo and Desvelo near to Riberalta	University Beni
Valera, Huanger	Social participation in Community Forestry projects, Bolivia	University Beni
van Heeswijk, Laura	Decentralization of Forest Management in Lowland Bolivia: The "Reality" of Decentralization Processes and the Possibilities for Local Communities	University of Wageningen
ASA student group	Commercial potential of Non-Timber-Forest Products from the Bolivian Amazon on German markets	University Beni & University Freiburg

References

Acosta, A. and Falconí F. (eds) (2005). *Asedios a lo imposible: propuestas económicas en construcción*, Instituto Latinoamericano de Investigaciones Sociales (ILDES), Quito

Aghón, G., Alburquerque, F. and Cortés, P. (2001). *Desarrollo económico local y descentralización en América Latina: un análises comparativo*, Instituto Latinoamericano y del Caribe de Planificación Económica y Social (ILPES)/Comisión Económica para América Latina y el Caribe (CEPAL), Santiago de Chile

Albuquerque, F. (2001). *Desenvolvimento económico local: caminhos e desafios para a construção de uma nova agenda política*, Banco Nacional de Desenvolvimento (BNDES), Rio de Janeiro

Aldrich, S. P., Walker, R. T., Arima, E. Y. and Caldas, M. M. (2006). 'Land-cover and landuse change in the Brazilian Amazon: Smallholders, ranchers and frontier stratification', *Economic Georgraphy*, vol 82, pp.265–88

Alexiades, M. N. (1999). 'Ethnobotany of the Ese Eja: plants, health and change in an Amazonian society', PhD thesis, City University of New York

Almeida, E., Sabogal, C. and Breinza, S. (2006). *Recuperação de áreas alteradas na Amazônia Brasileira: experiências locais, lições aprendidas e implicações para políticas públicas*, Center for International Forestry Research (CIFOR), Bogor

Altenburg, T. and Meyer-Stamer, J. (1999). 'How to promote clusters: Policy experiences from Latin America', *World Development*, vol 27, no 9, pp.1693–713

Alwang, J. and Siegel, P. B. (2003). 'Measuring the impacts of agricultural research on poverty reduction', Agricultural Economics, vol 29, pp.1–14

Amaral P. and Amaral Neto M. (2005). *Manejo florestal comunitário: processos e aprendizagens na Amazônia brasileira e na América Latina*, Instituto Internacional de Educação do Brasil (IEB)/Instituto do Homem e Meio Ambiente da Amazônia (IMAZON), Belém, Brazil

Andersen, L. (2007). 'The influence of land use on the soil fertility in the Ecuadorian Amazon', Degree dissertation, University of Freiburg

Angelsen, A. and Wunder, S. (2003). 'Exploring the forest – poverty link: key concepts, issues and research implications', Occasional Paper 40, Center for International Forestry Research, Bogor

Archer, M. (1995). *Realist social theory: the morphogenetic approach*, Cambridge University Press, Cambridge

Arias, G., Villacréz, D., Leon, M. and Andrade, M. (2006). *Proceso de diálogo nacional sobre control forestal y acuerdo intersectorial para un sistema nacional de control forestal*, Ministry of Environment, San José de Puembo, Ecuador

Arima, E., Schneider, R. R., Veríssimo, A., Souza, C. and Barreto, P. (2002). *Sustainable Amazon: Limitations and opportunities for rural development*, World Bank, Washington D.C.

Arroyo, M. G., Caldart, R. and Molina, M. C. (2004). *Por uma educação do campo*, Vozes, Petrópolis, Brazil

Arts, B. and Buizer, M. (2009). 'Forests, discourses, institutions. A discursive-institutional analysis of global forest governance', *Forest Policy and Economics*, vol 11, pp.340–7

Ashby, J. A., Braun, A. R., Gracia, T., Guerrero, M. P., Hernández, L. A., Quirós, C. A. and Roa, J. I. (2000). *Investing in farmers as researchers: Experience with local agricultural research committees in Latin America* Centro Internacional de Agricultura Tropical (CIAT), Cali, Colombia

Asner, P. G., Knapp, D. E., Broadbent, E. N., Oliveira, P. J. C., Keller, M. and Silva. J. N. (2005). 'Selective logging in the Brazilian Amazon', *Science*, vol 310, no 5747, pp.480–2

Assies, W. (1997). *Going nuts for the rainforest: Non-timber forest products, forest conservation and sustainability in Amazonia*, Thela, Amsterdam

Barclay, F., Rodríguez, M., Santos, F. and Valcárcel, M. (1991). *Amazonia 1940-1990: el extravío de una ilusión*, Terra Nuova & PUCP, Lima

Bardhan, P. (2000). 'Understanding underdevelopment: challenges for institutional economics from the point of view of poor countries', *Journal of Institutional and Theoretical Economics*, vol 156, pp.217–35

Balee, W. and Erickson, L.C. (eds) (2006). *Time and complexity in historical ecology: studies in the neotropical lowlands*, Columbia University Press, New York

Barreto, P., Pinto, A., Brito, B. and Hayashi, S. (2008). *Quem é o dono da Amazônia? Uma análise do recadastramento de imóveis rurais.* Instituto do Homem e Meio Ambiente da Amazônia (IMAZON), Belém, Brazil

Bebbington, A. (1999). 'Capitals and capabilities: a framework for analyzing peasant viability, rural livelihoods and poverty', *World Development*, vol 27, no 12, pp.2021–44

Becker, B. K. (2005). 'Geopolítica da Amazônia', *Estudos Avançados*, vol 19, no 53, pp.71–86

Becker, C. D. and Ostrom, E. (1995). 'Human ecology and resource sustainability: the importance of institutional diversity', *Annual Review Ecological Systems*, vol 26, pp.113–33

Belcher, B. and Schreckenberg, K. (2007).'Commercialisation of non-timber forest products: A reality check', *Development Policy Review*, vol 25, no 3, pp.355–77

Benatti, J. H., Santos, R. A. and da Gama, A. S .P. (2008). 'A grilagem das terras públicas na Amazônia brasileira', Série Estudos, Ministry of Environment, Brasília, Brazil

Benneker, C. (2008). 'Dealing with the State, the market and NGOs: The impact of institutions on the constitution and performance of Community Forest Enterprises (CFE) in the lowlands of Bolivia', PhD thesis, University of Wageningen

Biedenweg, K. (2009). 'El rol de los proyectos de apoyo en el Manejo Forestal Comunitario Pando, Bolivia', Report, University of Florida, Gainsville

Biggs, S. (1990). 'A multiple source of innovation model of agricultural research and technology promotion', *World Development*, vol 18, no 11, pp.1481–99

Billard, C. (2008). 'The influence of social and economic parameters on landscape dynamics and land use changes in the Amazon at the household-level using remotely-sensed imagery. A case study in the community of Palmares II, Brazil', MSc thesis, University of Freiburg

Binswanger, H. P. (1991). 'Brazilian policies that encourage deforestation in the Amazon', *World Development*, vol 19, no 7, pp.821–9

Blaikie, P. (1999). 'A review of political ecology: Issues, epistemology and analytical narratives', *Zeitschrift für Wirtschaftsgeographie*, vol 43, no 3–4, pp.131–47

Blum, B. (2009). 'Comparative study of development paradigms in the Bolivian, Peruvian, Ecuadorian and Brazilian Amazon', Report, University of Freiburg

Brandão Jr., A. and Souza Jr., C. (2006). 'Desmatamento nos assentamentos de reforma agrária na Amazônia', O Estado da Amazônia, vol 7, Instituto do Homem e Meio Ambiente da Amazônia (IMAZON), Belém, Brazil

Brazilian Ministry of Finance (2008). 'The country is advancing', Sustainable Economy, vol. 2, Editora Abril S.A, Brasilia, Brazil

—(2009). 'Back to growth', Sustainable Economy, vol 5, Editora Abril S.A., Brasilia, Brazil

Brazil's National Press (2005). 'Lei n°11.284/06', Diário Oficial da União, O livro branco da grilagem de terras no Brasil, Instituto Nacional da Colonização e Reforma Agrária (INCRA), Brasília, Brazil

Brito, B. and Barreto, P. (2006). 'A eficácia da aplicação da lei de crimes ambientais pelo IBAMA para proteção de florestas no Pará', *Revista de Direito Ambiental, Ed. Revista dos Tribunais*, vol 43, pp.35–65

Brondizio, E. S. (2004). 'Agriculture intensification, economic identity, and shared invisibility in Amazonian peasantry: Caboclos and colonists in comparative perspective', Culture and Agriculture, vol 26, no 1–2, pp.1–24

Browder, J., Pedlowski, M. and Summers, P. (2004). 'Land use patterns in the Brazilian Amazon: Comparative farm-level evidence from Rondônia', *Human Ecology*, vol 32, no 2, pp.197–224

Browder, J. O., Pedlowski, M. A., Walker, R., Wynne, R. H., Summers, P. M., Abad, A., Becerra-Cordoba, N., and Mil-Homens, J. (2008). 'Revisiting theories of frontier expansion in the Brazilian Amazon: A survey of the colonist farming population in Rondônia's post-frontier, 1992-2002', *World Development*, vol 36, no 8, pp1469–92

Brown, R., Stephens, C., Ouma, J., Murithi, M. and Barrett, C. (2006). 'Livelihood strategies in the rural Kenyan highlands', *African Journal of Agricultural Resource Economics*, vol 1, no 1, pp.21–36.

Bryant, L. R. (1998). 'Power, knowledge and political ecology in the third world: a review', *Progress in Physical Geography*, vol 22, no 1, pp.79–94

Bunker, S. (1985). *Underdeveloping the Amazon: extraction, unequal exchange, and the failure of the modern state*, University of Illinois Press, Champaign

Cáceres, A. L. E. (2005). *Análisis e impactos de la carretera Interoceanica*, Programa Competitividad, Innovación, Desarrollo (CID), Arequipa, Peru

Campos, P. (2009). 'Societal change in Amazonian communities as a consequence of the intervention of development organisations: Results from the study of the indigenous community of Callería in Peru', MSc thesis, University of Freiburg

Campos, J. J., Finegan, B. and Villalobos, R. (2001). 'Management of goods and services from neotropical forest biodiversity: diversified forest management in Mesoamerica' in Secretariat of the Convention on Biological Diversity (SCBD) 'Assessment, conservation and sustainable use of forest biodiversity', CBD Technical Series, vol 3, SCBD, Montreal, pp.5–16

Campos, M. T. and Nepstad, D. C. (2006). 'Smallholders, the Amazon's new conservationists', *Conservation Biology*, vol 20, no 5, pp.1553–6

Cano, W. (2012). 'Formal institutions, local arrangements and conflicts in the northern Bolivian communities after forest governance reforms', PhD thesis, Scientific Series, vol 15, Utrecht University

Carrero, G. C. and Fearnside, P. M. (2011). 'Forest clearing dynamics and the expansion of landholdings in Apuí, a deforestation hotspot on Brazil's Transamazon Highway', *Ecology and Society*, vol 16, no 2: 26, [online] http://www.ecologyandsociety.org/vol16/iss2/art26/

Carvalheiro, K., Sabogal, C. and Amaral, P. (2008). *Análise da legislação para o manejo florestal por produtores de pequeña escala na Amazônia brasileira.* Centre for International Forestry Research (CIFOR), Belém, Brazil

Cavendish, W. (1999). 'Poverty, inequality and environmental resources: quantitative analysis of rural households', Working Paper Series, vol 99–9, Center for the Study of African Economics, Oxford

CEETEPS (Centro Estadual de Educação Tecnológica Paula Souza) (2000) *Retrato falado da alternância: sustentando o desenvolvimento rural através da educação.* CEETEPS, São Paulo

Chambers, R. (1987). 'Sustainable livelihoods, environment and development: putting poor rural people first', Discussion Paper, vol 240, Institute of Development Studies (IDS), Brighton

Chambers, R. and Conway, G. R. (1992). 'Sustainable rural livelihoods: practical concepts for the 21st century', Discussion Paper, vol 296, Institute of Development Studies (IDS), Brighton

Chapin, M. (2004). 'A challenge to conservationists', *World Watch Magazine*, 11 Dec. 2004, UNO/BOL/723/DCP.

Chavez, P. (1996). 'Image-based atmospheric corrections revisited and improved', *Photogrammetric Engineering and Remote Sensing*, vol 62, no 9, pp.1025–36

Chhatre, A. and Agrawal, A. (2009). 'Trade-offs and synergies between carbon storage and livelihood benefits from forest commons', *Proceedings of the National Academy of Sciences of the United States of Amercia (PNAS)*, vol 106, pp.17667–70

Chirif, A. and García Hierro, P. (2007). *Marcando* territorio: *Progresos y limitaciones de la titulación de* territorios *indígenas en la Amazonía*, International Working Group for Indigenous Affairs, Copenhagen

Chomitz, K. and Thomas, T. (2003). 'Determinants of land use in Amazonia: A fine-scale spatial analysis', *American Journal of Agricultural Economics*, vol 85, no 4, pp.1016–28

Chomitz, K. M., Buys, P., De Luca, G., Thomas, T. S. and Wertz-Kanounnikoff, S. (2006). 'At loggerheads? Agricultural expansion, poverty reduction and environment in the tropical forests', World Bank Policy Research Report, Development Research Group, World Bank, Washington D.C.

Cleaver, F. (2002). 'Reinventing institutions: bricolage and social embeddedness of natural resource management', *The European Journal of Development Research*, vol 14, no 2, pp.11–30 (20)

Cornia, G. A. (1985) 'Farm size, land yields and the agricultural production function: An analysis for 15 developing countries', *World Development*, vol 13, no 4, pp.513–34

Cronkleton, P., Guariguata R. M. and Albornoz, A. M. (2012). 'Multiple use forestry planning: Timber and Brazil nut management in the community forests of Northern Bolivia', *Forest Ecology and Management*, vol 268, no 1, pp.49–56

Current, D. and Scherr, S. J. (1995). 'Farmer costs and benefits from agroforestry and farm forestry projects in Central America and the Caribbean: implications for policy', *Agroforestry Systems*, vol 30, no 1, pp.87–103

D'Antona, A., VanWey, L. and Hayashi, C. (2006). 'Property size and land cover change in the Brazilian Amazon', *Population and Environment*, vol 27, no 5–6, pp.373–96

Davies, S. and Hossain, N. (1997). *Livelihood adaptation, public action and civil society: a review of the literature*, Institute of Development Studies (IDS), Brighton

Dean, W. (1989). *A luta pela borracha no Brasil: um estudo de história ecológica*, Nobel, São Paulo

Deetz, S. and Radford, G. (2008). *Communication theory at the crossroads: Theorizing for globalization, pluralism and collaborative needs*, Blackwell Publications, Oxford

Defensoria del Pueblo del Peru (2010) *Reporte Conflictos Sociales*, vol 72, Lima.

De Groot S. R., Wilson, A. M. and Boumans, M. J. R. (2002). 'A typology for the classification, description and valuation of ecosystem functions, goods and services', *Ecological Economics*, vol 41, pp.393–408

De Jong, W., Ruiz, S. and Becker, M. (2006). 'Conflicts and communal forest management in northern Bolivia', *Forest Policy and Economics*, vol 8, pp.447–457

De Koning, J. (2011) 'Reshaping institutions: bricolage processes in smallholder forestry in the Amazon', PhD thesis, University of Wageningen

De Schutter, O. (2011). 'How not to think of land-grabbing. Three crititques of large-scale investments in farmland', *Peasant Studies Journal*, vol 38, no 2, pp.249–79

Denevan, W. (1992). 'The pristine myth: The landscape of the Americas in 1492', *Annals of the American Association of Geographers*, vol 82, no 3, pp.369–85

Depzinski, T. (2007). 'Eligible local partners of development organisations: criteria of development organisations for selecting local partners: A study in the Peruvian Amazon', Degree dissertation, University of Göttingen, Germany

Dethier, J. J., Pestieau, P. and Ali, R. (2010). 'Universal minimum old age pensions. Impact on poverty and fiscal cost in Latin American countries', Policy Research Working Paper, vol 5292, World Bank, Washington D.C.

DFID (UK Department for International Development) (1999). *Sustainable livelihoods guidance sheets*, DFID, London, [online] www.livelihoods.org

Di Sabatto, A. (2001). *Perfil dos proprietários / detentores de grandes imóveis rurais que não atenderam à notificação da portaria 558/99*. Instituto Nacional da Colonização e Reforma Agrária (INCRA), Brasília, Brazil

Donovan, J., Stoian, D. and Poole, N. (2008). 'A global review of rural community enterprises', Rural Enterprise Development Collection, vol 2, Tropical Agricultural Research and Higher Education Center (CATIE), Turialba, Costa Rica

Du, Y., Teillet, P. and Cihlar, J. (2002). 'Radiometric normalization of multitemporal high-resolution satellite images with quality control for land cover change detection', *Remote Sensing of Environment*, vol 82, no 1, pp.123–34

Earthjustice (2007). *Environmental rights report 2007*, Earthjustice, Oakland

Ehringhaus, C. (2005). 'Post-victory dilemmas: Land use, development policies, and social movement in Amazonian Extractive Reserves', PhD thesis, Yale University School of Forestry and Environmental Studies

EFTRN (European Tropical Forest Research Network) (2004). 'National Forest Programmes' *EFTRN News*, vol 41–42

Elder-Vass, D. (2007). 'A method for social ontology: iterating ontology and social research' *Journal of Critical Realism*, vol 6, no 2, pp.226–49

Ellis, F. (2000). *Rural livelihoods and diversity in developing countries*, Oxford University Press, Oxford

Engel, S., Pagiola, S. and Wunder, S. (2008). 'Designing payments for environmental services in theory and practice: An overview of the issues', *Ecological Economics*, vol 65, no 4, pp.663–74

Escalera, E. and Oporto, T. (in elaboration) 'Proceso de adopción de sistemas agroforestales en la comunidad campesina Palmira del Norte Amazónico de Bolivia', in B. Pokorny, I. Montero, J. C. Montero and J. Johnson (eds) 'Uso forestal por pequeños productores en la Amazonía: En busca de evidencias empíricas para los grandes paradigmas', University of Freiburg

Fairhead, J., Leach, M., and Scoones, I. (2012). 'Green grabbing: a new appropriation of nature?', *The Journal of Peasant Studies*, vol 39, no 2, pp.237–61

FAO (Food and Agriculture Organisation of the United Nations) (1992) 'Sociological analysis in agricultural investment project design' FAO Investment Centre Technical Paper, vol 9, FAO, Rome

—(2010). 'Global forest resources assessment 2010', Main report. FAO Forestry Paper, vol 163, FAO, Rome [online] http://www.fao.org/forestry/fra/fra2010/en/

Farrington, J., Carney, D., Ashley, C. and Turton, C. (1999). 'Sustainable livelihoods in practice: early applications of concepts in rural areas', Natural Resource Perspective, vol 42, Overseas Development Institute (ODI), London

Fearnside, P. M. (1993). 'Deforestation in the Brazilian Amazon: the effect of population and land tenure', Ambio, vol 22, no 8, pp.537–545

—(2005). 'Deforestation in Brazilian Amazonia: History, rates, and consequences', *Conservation Biology*, vol 19, pp.680–688

—(2008). 'The roles and movements of actors in the deforestation of Brazilian Amazonia', *Ecology and Society*, vol 13, no 23 [online] http://www.ibcperu.org/doc/isis/9004.pdf

Ferreyros, S. J. P. and Medina, G. (in preparation). 'Motivaciones para el cambio en el uso del suelo por los campesinos en áreas peri-urbanas de la Amazonía Peruana', in B. Pokorny, I. Montero, J. C. Montero and J. Johnson (eds) 'Uso forestal por pequeños productores en la Amazonía: En busca de evidencias empíricas para los grandes paradigmas', University of Freiburg

Ferraz, S. F. B., Vettorazzi, C. A. and Theobald, D. M. (2009). 'Using indicators of deforestation and land-use dynamics to support conservation strategies: a case study of central Rondônia, Brazil', *Forest Ecology and Management*, vol 257, pp.1586–95

Flick, K. (2008). 'Clarifying the understanding of learning and participation: an exploratory survey of participatory forestry approaches in the Amazon', MSc thesis, University of Freiburg

Freire, P. (1979). *Educação e mudança*, 16th edn, Paz e Terra, Rio de Janeiro

—(1983). *Extensão ou comunicação?*, Paz e Terra, Rio de Janeiro

FSC-AC (The Forest Stewardship Council – Asociación Civil) (2010). *Global FSC certificates, types and distribution.* FSC, Bonn

Fuentes, D., Haches, R., Maldonado, R., Albornoz, M., Cronkleton, P., de Jong, W. and Becker, M. (2005). *Pobreza, descentralización y bosque en la Amazonia Boliviana*, Center for International Forestry Research (CIFOR), Bogor

Fujisaka, S. and White, D. (1998). 'Pasture or permanent crops after slash-and-burn cultivation? Land-use choice in three Amazon colonies', *Agroforestry Systems*, vol 42, pp.45–59

Gallagher, K. (1999). *Las escuelas del campo para agricultures (ECA). Un proceso de extension grupal basado en métodos de educación no formal para adultos.* FAO, Global IPM Facility, Rome

Garcia, E.; Montero, J. C. and Pokorny, B. (in preparation) 'La compatibilidad de la adopción de productos con alta demanda comercial por pequeños productores con su demanda por seguridad alimentaria: El caso del camu-camu (Myrciaria dubia HBK Mc Vaugh) en Yarinacocha, Perú', in B. Pokorny, I. Montero, J. C. Montero and J. Johnson (eds) 'Uso forestal por pequeños productores en la Amazonía: En busca de evidencias empíricas para los grandes paradigmas', University of Freiburg

Garnett, S., Sayer, J. and Du Toit, J. (2007). 'Improving the effectiveness of interventions to balance conservation and development: a conceptual framework', *Ecology and Society*, vol 12, no 1, 2, [online] http://www.ecologyandsociety.org/vol12/iss1/art2/

Gasche, J. (ed.) (2004). *Crítica de proyectos y proyectos críticos de desarrollo: Una reflexión latinoamericana con énfasis en la Amazonia*, Instituto de Investigaciones de la Amazonía Peruana (IIAP), Iquitos

Gasche, J. and Vela M. N. (2012). *Sociedad bosquesina Vol. 1. Ensayo de antropología rural amazónica, acompañado de una crítica y propuesta alternativa de proyectos de desarrollo,* Instituto de Investigaciones de la Amazonía Peruana (IIAP), Iquitos

Gatter, S. and Romero, M. R. (2005). *Análisis económico de la cadena de aprovechamiento, transformación y comercialización de madera aserrada provenientes de bosques nativos en la región centro-sur de la amazonía ecuatoriana,* Servicio Forestal Amazônico (SFA), Macas, Ecuador

Ghazoul, J. and Sheil, D. (2010). *Tropical rain forest ecology, diversity, and conservation,* Oxford University Press, New York

Gibson, C. C., McKean, M. A. and Ostrom, E. (eds) (2000). *People and forests. Communities, institutions and governance.* MIT Press, Cambridge

Giugale, M. M.; Fretes-Cibils, V. and Lopez Calix, J. R. (eds) (2003). *Ecuador – An economic and social agenda in the new millennium.* World Bank, Washington D.C.

Godar, J. (2009). 'The environmental and human dimensions of frontier expansion at the Transamazon highway colonization area', PhD thesis, University of León, Spain

Godar, J., Tizado, J. E. and Pokorny, B. (2008). 'A expansão da fronteira na Transamazônica: o impacto comparado da agricultura familiar e da pecuária', Policy Brief, University of Freiburg

Godar, J., Tizado, J. E., Pokorny, B. (2012a). 'Who is responsible for deforestation in the Amazon? A spatially explicit analysis along the Transamazon Highway in Brazil', *Forest, Ecology and Management,* vol 267, pp.58–73

Godar, J., Tizado, J. E., Pokorny, B, Johnson, J. (2012b). 'Typology and characterization of Amazon colonists: A case study along the Transamazon Highway', *Human Ecology,* vol 40, no 2, pp.251–67

Government of Ecuador (1997). 'Ley de descentralización del Estado y de participación social', Ley No 27, Registro Oficial No 169, 8 October

Government of the State of Pará (2008). 'Um bilhão de árvores para Amazônia', Programa estadual de restauração florestal, Belém, Brazil

Grau, H. R. and Aide, M. (2008). 'Globalization and land use transitions in Latin America', *Ecology and Society,* vol 13, no 2, 1–12, [online] http://www.ecologyandsociety.org/vol13/iss2/art16/

Gray, J. (2002). 'Forest concession policies and revenue systems – country experience and policy changes for sustainable tropical forestry', World Bank Technical Paper, vol 522, World Bank, Washington D.C.

Green, D. (2008). *From poverty to power: How active citizens and effective states can change the world.* OXFAM International, Oxford

Green, M. and Hulme, D. (2005). 'From correlates and characteristics to causes: Thinking about poverty from a chronic poverty perspective', *World Development,* vol 33, no 6, pp.867–879

Greenpeace (2005). *Grilagem de terras na Amazônia: Negócio bilionário ameaça a floresta e populações tradicionais.* [online] http://www.greenpeace.org/brasil/PageFiles/3951/grilagem.pdf

Guijt, I. (ed.) (2007) *Negotiated learning: collaborative monitoring in forest resource management.* Resources for the Future, Washington D.C.

Hage, S. M. (2004). 'A concepção de educação do campo no cenário das políticas públicas da sociedade brasileira', Comunica Multissérie, vol 1, no 1, Edital, Belém.

Hall, A. (2008). 'Better RED than dead: paying the people for environmental services in Amazonia', Biological Sciences, vol 363, no 1498, pp.1925–32

Harriss, J. (2007). 'Bringing politics back into poverty analysis: Why understanding social relations matters more for policy on chronic poverty than measurement', CPRC Working Paper, vol 77, Chronic Poverty Research Centre, Manchester

Hazell, P., Poulton, C., Wiggins, S., Dorward, A. (2010). 'The future of small farms: Trajectories and policy priorities', *World Development*, vol 38, no 10, pp.1349–61

Hecht, S. (2005). 'Soybeans, development and conservation on the Amazon frontier', *Development and Change*, vol 36, no 2, pp.375–404.

—(2007). 'Globalization and forest recovery in inhabited landscapes', *Bioscience*, vol 57, no 8, pp.663–72.

—(2010). 'The new rurality: Globalization, peasants and the paradoxes of landscapes', *Land Use Policy*, vol 27, no 2, pp.161–9

—(2011). *The scramble for the Amazon and the "Lost Paradise" of Euclides da Cunha*, University of Chicago Press, Chicago

Hecht, S. and Cockburn, A. (1989). *The fate of the forest: Developers, destroyers, and defenders of the Amazon*. Harper Perennial, New York

Heltberg, R. (1998). 'Rural market imperfections and the farm size productivity relationship: Evidence from Pakistan', *World Development*, vol 26, no 10, pp.1807–26

Henkemans, A. (2001). 'Tranquilidad and hardship in the forest. Livelihood capitals and perceptions of Camba forest dwellers in the northern Bolivian Amazon', PhD thesis, University of Utrecht

Herrera, J. A. (2011). 'Desenvolvimento capitalista e realidade da produção agropecuária familiar na Amazônia Paraense', PhD thesis, UNICAMP, Campinas

Hoch, L. (2009). 'Do smallholders in the Amazon benefit from tree growing?', PhD thesis, University of Freiburg

Hoch, L., Pokorny, B. and De Jong, W. (2009). 'How successful is tree growing for smallholders in the Amazon?' *International Forestry Review*, vol 11, no 3, pp.299–310

—(2012). 'Financial attractiveness of smallholder tree plantations in the Amazon: bridging external expectations and local realities', *Agroforestry Systems*, vol 84, no 3, pp.361–75

Hoch, L., Pokorny, B. and Medina, G. (2008). 'Plantaciones forestales por productores familiares en la Amazonía', Policy Brief, University of Freiburg

Hodgson, G. M. (2000). *Structures and institutions: reflections on institutionalism, structuration theory and critical realism*, The Business School, University of Hertfordshire, England

Holt-Giménez, E. and Spang, L. (2005). 'IIRSA Update #1' Bank Information Centre, Washington D.C., 28 February, [online] http://www.bicusa.org/en/Article.1946.aspx

Homma, A. K. O. (2006). 'Agricultura familiar na Amazônia: a modernização da agricultura itinerante', in I. S. F. Sousa (ed.) *Agricultura familiar na dinâmica da pesquisa agropecuária*. Embrapa Informação Tecnológica, Brasília, Brazil, pp.37–60

Humphreys, D. (2008). 'The politics of avoided deforestation': historical context and contemporary issues', *International Forestry Review*, vol 10, no 3, pp.433–42

Humphries, S. S. and Kainer, K. A. (2006). 'Local perceptions of forest certification for community-based enterprises', *Forest Ecology and Management*, vol 235, no 1–3, pp.30–43

Ibarra, E., Romero, M. and Gatter, S. (2008). *Análisis del marco legal para el manejo forestal por pequeños productores rurales en la Amazonia ecuatoriana*, Servicio Forestal Amazônico (SFA), Macas, Ecuador

IBGE (Instituto Brasileiro de Geografia e Estatística) (2009). 'Uso da terra e a gestão do território no Estado do Acre', Informe Técnico

IDB (Interamerican Development Bank) (2006). *Construyendo un nuevo continente. Un enfoque regional para fortalecer la infraestructura sulamericana*, IDB, Washington D.C.

IMAZON (Instituto do Homem e Meio Ambiente da Amazônia) (2005). *A destinação dos bens apreendidos em crimes ambientais no Pará*, IMAZON, Belém, Brazil

Ingles, A., Musch, A., and Qwist-Hoffman, H. (1999). *The participatory process for supporting collaborative management of natural resources: an overview*, FAO. Rome

IPAM (Instituto de Pesquisa Ambiental na Amazônia) (2006). 'A grilagem de terras públicas na Amazônia brasileira', Série Estudos, Ministry of Environment, Brasília, Brazil

IPHAE (Instituto para el Hombre Agricultura y Ecología) (2007). 'Diagnóstico rural participativo campo central', Report, IPHAE, Riberalta, Bolivia

ITERPA (Instituto de Terras do Pará) (2007). *Iterpa e o ordenamento territorial o Estado do Pará. A regularização fundiária como instrumento de ordenar o espaço e democratizar o acesso à terra.* ITERPA, Belém, Brazil

Jollivet, M. (ed.) (1974). *Les collectivités rurales françaises, tome II: Sociétés paysannes ou lutte des classes au village?* Armand Colin, Paris

Jollivet, M. and Mendras, H. (1971). *Les collectivités rurales françaises, tome 1.* Armand Colin, Paris

Jonasse, R. (2009). *Agrofuels in the Americas.* Institute for Food and Development Policy, Oakland

Kaag, M., Van Berkel, R., Bros, J., De Bruijn, M., Van Dijk, H., De Haan, L., Nooteboom, G. and Zoomers, A. (2004). 'Ways forward in livelihoods research', in D. Kalb, W. Pansters and H. Siebers (eds) *Globalization and development. Themes and concepts in current research,* Kluwer Academic Publishers, Dordrecht, pp.49–74

Kaimowitz, D. (2003). 'Forest law enforcement and rural livelihoods', *International Forestry Review,* vol 5, no 3, pp.199–210

Keen, B. and Haynes, K. (2009). *A history of Latin America,* 8th edn, Houghton Mifflin Company, Boston

Kern, M. (2012). 'Uncovering local markets for Amazonian smallholders: The case of the Extractive Reserve "Verde para Sempre" in Porto de Moz in Brazil', MSc thesis, University of Freiburg

Killeen, T. J. (2007). 'A perfect storm in the Amazon wilderness: development and conservation in the context of the initiative for the integration of the regional infra-structure of South America (IIRSA)', Report, Conservation International (CI), Arlington

Klein, H. S. (2011). *A concise history of Bolivia,* 2nd edn, Stanford University, USA

Kohlhepp, G. (1991). 'The destruction of the tropical rain forests in the Amazon region of Brazil. An analysis of the causes and the current situation', *Applied Geography and Development,* vol 38, pp.87–109

Lal, P., Lim-Applegate, H. and Scoccimarro, M. (2001). 'The adaptive decision-making process as a tool for integrated natural resource management: focus, attitudes, and approach', *Conservation Ecology,* vol 5, no 2, 11, [online] http://www.consecol.org/vol5/iss2/art11

Lamarche, H. (ed.) (1993). *A agricultura familiar.* Editora Unicamp, Campinas, Brazil

Lambin, E. F. and Meyfroidt, P. (2010). 'Land use transitions: Socio-ecological feedback versus socio-economic change', *Land Use Policy,* vol 27, no 2, pp.108–18

Lange, K. (2008). 'Contribution of medical plants to the primary health care of indigenous communities; Case study of two Shipibo-Conbio communities in the Ucayali region, Peru', BSc thesis, University of Freiburg

Larson, A., Barry, D., Dahal, R. G. and Colfer, C. (eds) (2010). *Forests for people: Community rights and forest tenure reform.* Earthscan, London

Larson, A., Cronkleton, P., Barry, D. and Pacheco, P. (2008). 'Tenure rights and beyond: Community access to forest resources in Latin America', Occasional Paper, vol 50, Centre for International Forestry Research (CIFOR), Bogor

Larson, A., Pacheco, P., Toni, F. and Vallejo, M. (2006). *Exclusión e inclusión en la forestería Latinoamérica: Hacia dónde va la descentralización?',* Centro Internacional de Investigaciones para el Desarollo (IDRC), La Paz

Larson, A. and Ribot, J. (2004). 'Democratic decentralization through a natural resource lens: An introduction', *The European Journal of Development Research*, vol 16, no 1, 1–25

Larson, A. and Soto, F. (2008). 'Decentralization of natural resource governance regimes', *Annual Review of Environment and Resources*, vol 33, pp.213–239, [online] 10.1146/annurev.environ.33.020607.095522

Laurance, W. F., Delamonica, P., Laurance, S. G., Vasconcelos, H. L. and Lovejoy, T. E. (2000). 'Conservation: Rainforest fragmentation kills big trees', *Nature*, vol 404, p.836

Laurance, W. F., Laurance, S. G. and Delamonica, P. (1998). 'Tropical forest fragmentation and greenhouse gas emissions', *Forest Ecology and Management*, vol 110, no 1–3, pp.173–180.

Laurance, W. F., Laurance, S. G., Ferreira, V. F., Rankin-de Merona, J. M., Gaston, C. and Lovejoy, T. E. (1997). 'Biomass collapse in Amazonian forest fragments', *Science*, vol 278, pp.1117–18.

Laurance, W. F., Lovejoy, T. E., Vasconcelos, H. L., Bruna, E. M., Didham, R. K., Stouffer, P. C., Gascon, C., Bierregaard, R. O., Laurance, S. G. and Sampaio, E. (2002). 'Ecosystem decay of Amazonian forest fragments: A 22-year investigation', *Conservation Biology*, vol 16, pp.605–18

Lawlor, K. and Huberman, D. (2009). 'Reduced emissions from deforestation and forest degradation (REDD) and human rights', in J. Campese, T. Sunderland, T. Greiber and G. Oviedo (eds) *Rights-based approaches: Exploring issues and opportunities for conservation*, Centre for International Forestry Research (CIFOR), Bogor, pp.269–85

Lentini, M., Pereira, D., Celentano, D. and Pereira, R. (2005). *Fatos florestais da Amazônia 2005*, Instituto do Homem e Meio Ambiente da Amazônia (IMAZON), Belém, Brazil

Lima, E., Leite, A., Nepstad, D., Kalif, K., Ramos, A. and Pereira, S. (2003). *Florestas familiares: um pacto sócio-ambiental entre a indústria madeireira e agricultura familiar na Amazônia*. Instituto de Pesquisa Ambiental Amazônia (IPAM), Belém, Brazil

Lindenmayer, D. B., Cunningham, R. B., Donnelly, C. F. and Lesslie, R. (2002). 'On the use of landscape surrogates as ecological indicators in fragmented forests', *Forest Ecology and Management*, vol 159, no 3, pp.203–16

Long, N. and Van der Ploeg, J. D. (1989). 'Demythologizing planned interventions: an actor perspective', *Sociologia Ruralis*, vol 29, no 3/4, pp.226–249

Lu, D. (2005). 'Integration of vegetation inventory data and Landsat TM image for vegetation classification in the western Brazilian Amazon', *Forest Ecology and Management*, vol 213, no 1-3, pp.369–83

Lu, D., Mausel, P., Brondizio, E. and Moran, E. (2003). 'Classification of successional forest stages in the Brazilian Amazon basin', *Forest Ecology and Management*, vol 181, no 3, pp.301–12

MA (Millennium Ecosystem Assessment) (2005). *Ecosystems and human well-being: synthesis*. Island Press, Washington, D.C.

Macqueen, D. (ed) (2012). *Supporting small forest enterprises – A facilitator's toolkit. Pocket guidance not rocket science!*, International Institute for Environment and Development (IIED), Edinburgh

MAE (Ecuadorian Ministy of Environment) (2006a). *Sistema nacional descentralizado de control forestal (SNDCF)*. MAE, Quito, Ecuador

—(2006b). 'Plan Nacional de Forestación y Reforestación', Acuerdo Ministerial 113, MAE, Quito, Ecuador

März, D. (2008). 'Effects of road construction on a long-term agro-forestry initiative in the region of Puerto Maldonado in the Peruvian Amazon', BSc thesis, University of Freiburg

Manel, S., Schwartz, M. K., Luikart, G. and Taberlet, P. (2003). 'Landscape genetics: combining landscape ecology and population genetics', *Trends Ecological Evolution*, vol 18, pp.189–97

Mann, C. C. (2005) *1491: New revelations of the Americas before Columbus*. Knopf Publishing Group, New York

Marshall, M. N. (1996). 'Sampling for qualitative research', *Family Practice*, vol 13, pp.522–5

Margulis, S. (2003). 'Causes of deforestation of the Brazilian Amazon', World Bank Working Paper Series, vol 100, World Bank, Washington D.C.

Martín-Barbero, J. (2006). *Dos meios às mediações: comunicação, cultura e hegemonia*, 4th edn, Universidad Federal do Rio de Janeiro (UFRJ)

Martins, H., Amaral, P., Nascimento, K. and Reis, R. (2007). 'Avaliação da pressão humana na Reserva Extrativista Verde para Sempre no oeste do Pará', Anais XIII, Simpósio Brasileiro de Sensoriamento Remoto, Florianópolis, Brasil, 21–26 abril 2007, Instituto Nacional de Estudos Espaciais (INPE), São José dos Campos, Brazil, pp.2817–24

Martínez Montaño, J. A. (2008). *Marco legal para el manejo forestal por pequeños productores y comunidades en las tierras bajas de Bolivia*. Centre for International Forestry Research (CIFOR), Santa Cruz, Bolivia

Masias, K. (2011). 'Challenges of company-community partnerships within REDD+: A case study of Brazil nut concessions in Madre de Dios, Peru', MSc thesis, University of Freiburg

McKean, M. A. (2000). 'Common property: what is it, what is it good for, and what makes it work', in C. C. Gibson, M. A. McKean and E. Ostrom (eds) *People and forests. Communities, institutions and governance*. MIT Press, Cambridge, pp.27–55

McMichael, P. (ed.) (2010). Contesting development. Critical struggles for social change, Routledge, New York

Medina, G. (2008). 'Moving from dependency to autonomy: an opportunity for local communities in the Amazon frontier to benefit from the use of their forests', PhD thesis, University of Freiburg

Medina, G. and Shanley, P. (2004). 'Big trees, small favors: loggers and communities in Amazonia', *Bois et Forets des Tropiques*, vol 280, no 4, pp.19–25

Medina, G. and Pokorny, B. (2008). *Avaliação financeira do manejo florestal comunitário*. Instituto Brasileiro do Meio Ambiente e dos Recursos Naturais Renováveis (IBAMA), Brasilia, Brazil

—(2011). 'Avaliação financeira do manejo florestal comunitário', *Novos Cadernos NAEA*, vol 14, no 2, pp.25–36

Medina, G., Pokorny, B. and Campbell, B. (2008). 'Favoreciendo el desarrollo local en la Amazonia: lecciones de las iniciativas de manejo forestal comunitario', Forest Livelihood Brief 8, Center for International Forestry Research (CIFOR), Bogor

Medina, G., Pokorny, B. and Weigelt, J. (2009a). 'The power of discourse: hard lessons for traditional forest communities in the Amazon', *Forest Policy and Economics*, vol 11, pp.392–7

Medina, G., Pokorny, B. and Campbell, B. (2009b). 'Loggers and development agents exercising power over Amazonian villagers', *Development and Change*, vol 40, no 4, pp.745–67

Medina, G., Pokorny, B. and Campbell, B. (2009c). 'Community forest management for timber extraction in the Amazon frontier', *International Forestry Review*, vol 11, no 3, pp.408–20

Mertens, B., Poccard-Chapuis, R., Piketty, M. G., Lacques, A. E. and Venturieri, A. (2002).

'Crossing spatial analyses and livestock economics to understand deforestation processes in the Brazilian Amazon: the case of Sao Felix do Xingu in South Para', *Agricultural Economics*, vol 27, no 3, pp.269–94

Mesquita, R., Delamonica, P. and Laurance, W. F. (1999). 'Effects of surrounding vegetation on edge-related tree mortality in Amazonian forest fragments', *Biological Conservation*, vol 91, pp.129–34

Michalski, F., Metzger, J. P. and Peres, C.A. (2010). 'Rural property size drives patterns of upland and riparian forest retention in a tropical deforestation frontier', *Global Environmental Change*, vol 20, no 4, pp.705–12

Milani Filho, J., Gandin, C .L. and Cavalheiro, N. R. R. (2002). 'Casa familiar rural, a juventude aprendendo com a realidade', *Agropecuário Catarinense*, vol 15, no 2, pp.56–7

Milz, J. (1997). *Guía para el establecimiento de sistemas agroforestales*, 2nd edn, Deutscher Entwicklungsdienst (DED), La Paz

MINAG and INRENA (Peruvian Ministry for Agriculutre and the Instituto Nacional de Recursos Naturales) (2006). 'Plan Nacional de Reforestación (2005–2024)', Resolución Suprema 002-2006-AG, Lima

Miranda, P. (1990). 'Pensar extensionista: um caso de cegueira induzida. Preâmbulo para um estudo crítico da extensão rural no Estado do Pará', in C. M. Flores and T. A. Mitschein (eds) *Realidades amazônicas no fim do Século XX*. Asociación de las Universidades Amazónicas (UNAMAZ), Belém, Brazil, pp.365–431

MMA, MAPA, MDA andscott MCT (Ministry of Environment, Ministry of Transport, Ministry of Agrarian Development, Ministry of Science and Technology in Brazil) (2007). *Plano nacional de silvicultura com espécies nativas e sistemas agroflorestais (PENSAF)*, MMA, Brasilia, Brazil

Montero, J. C. (2007). 'From knowledge transfer to knowledge exchange: analysis of smallholders' perceptions on tree growing in the Amazon', MSc thesis, University of Freiburg

Moran, E. (1993). 'Deforestation and land use in the Brazilian Amazon', *Human Ecology*, vol 21, no 1, pp.1–21

Moran, E. F., Brondizio, E. S., Tucker, J. M., Da Silva-Forsberg, M. C., McCracken, S. and Falesi, I. (2000). 'Effects of soil fertility and land use on forest succession in Amazonia', *Forest Ecology and Management*, vol 139, pp.93–108

Morley, S. (1995). *Poverty and inequality in Latin America: the impact of adjustment and recovery in the 1980s*, The Johns Hopkins University Press, Baltimore

Myrdal, G. (1970). *The challenge of world poverty: a world anti-poverty program in outline*, Pantheon, New York

Nalvarte, J. (2011). 'Impacto del manejo forestal con fines maderables aplicado en la comunidad nativa Callería, región Ucayali-Perú', MSc thesis, Universidad Nacional Agraria La Molina, Lima

Nelson, A. and Chomitz, K. M. (2009). 'Protected area effectiveness in reducing tropical deforestation: a global analysis of the impact of protection status', Independent Evaluation Group Evaluation Brief, vol 7, World Bank, Washington D.C.

—(2011). 'Effectiveness of strict vs. multiple use protected areas in reducing tropical forest fires: A global analysis using matching methods', *PLOIS ONE*, vol 6, no 8, e22722

Nepstad, D., Azevedo-Ramos, C., Lima, E., McGrath, D., Pereira, C. and Merry, F. (2004). 'Managing the Amazon timber industry', *Conservation Biology*, vol 18, pp.1–3

Nepstad, D. and Schwartzman, S. (eds) (1992). Non-Timber products from tropical forests: Evaluation of a conservation and development strategy, New York Botanical Garden, New York

Nepstad, D., Stickler, C. and Almeida, O. (2006). 'Globalization of the Amazon soy and beef industries: opportunities for conservation', *Conservation Biology*, vol 20, pp.1595–603

Neumann, R. P. (2008). 'Probing the (in)compatibilities of social theory and policy relevance in Piers Blaikie's political ecology', *Geoforum*, vol 39, pp.728–35

Neumann, R. P. and Hirsch, E. (2000). *Commercialization of non-timber forest products: Review and analysis of research*, Centre for International Forestry Research (CIFOR), Bogor

Nygren, A. and Rikoon, S. (2008). 'Political ecology revisited: Integration of politics and ecology does matter', *Society and Natural Resources*, vol 21, pp767–82

O'Neill, J. (2001). 'Building better global economic BRICs', Global Economics Paper, vol 66, Goldman Sachs Global Research Centres, New York

Ortiz, S. (2007). 'Potenzial von Märkten für Waldprodukte von Kleinbauern: Ein Fallbeispiel aus Riberalta, Bolivien', MSc thesis, University of Freiburg

Ostrom, E. (1999). 'Self-governance and forest resources', Occasional Paper, vol 20, Centre for International Forestry Research (CIFOR), Bogor

Pacheco, P. (1992). *Integración económica y fragmentación social: el itinerario de las barracas en la Amazonia*, Centro de Estudios para el Desarrollo Laboral y Agrario (CEDLA), La Paz

—(2005). 'Populist and capitalist frontiers in the Amazon: Diverging dynamics of agrarian and landuse change', PhD thesis, Clark University, Worcester

—(2009a). 'Agrarian reform in the Brazilian Amazon: its implications for land distribution and deforestation', *World Development*, vol 37, no 8, pp.1337–47

—(2009b). 'Agrarian change, cattle ranching and deforestation: Assessing their linkages in Southern Pará', *Environment and History*, vol 15, pp.493–520 [online] 10.3197/0967340 09X12532652872072

Pacheco, P., Barry, D., Cronkleton, P., Larson, A. M. (2011). 'The recognition of forest rights in Latin America: progress and shortcomings of forest tenure reforms', *Society and Natural Resources*, ISSN: 0894-1920, [online] 10.1080/08941920.2011.574314

Pacheco, P., De Jong, W. and Johnson, J. (2010). 'The evolution of the timber sector in lowland Bolivia: Examining the influence of three disparate policy approaches', *Forest Policy and Economics*, vol 12, pp.271–76, [online] 10.1016/j.forpol.2009.12.002

Pacheco, P., Ibarra, E., Cronkleton, P. and Amaral, P. (2008). 'Politicas públicas que afectan el manejo forestal comunitario', in C. Sabogal, W. de Jong, B. Pokorny and B. Louman (eds) *Manejo forestal comunitario en América tropical: Experiencias, lecciones aprendidas y retos para el futuro*, Center for International Forestry Research (CIFOR), Bogor, pp.201–28

Pacheco, P., Nunes, W., Rocha, C., Vieira, I., Herrera, J. A., Dos Santos, K.A., Da Silva, T. F. and Cayres, G. (2009). *Acesso a terra e meios de vida: examinando suas interações em três locais no estado do Pará*. Center for International Forestry Research (CIFOR), Belém

Padoch, C., Brondizio, E., Costa, S., Pinedo-Vasquez, M., Sears, R. R. and Siquerira, A. (2008). 'Urban forest and rural cities: multi-sited households, consumption patterns, and forest resources in Amazonia', *Ecology and Society*, vol 13, no 2, [online] www.ecologyandsociety.org/vol13/iss2/art2/

Palm C. A., Vosti S. A., Sanchez P. A., Ericksen P. J. and Juo A. S. R. (eds) (2005). *Slash and burn: the search for alternatives*, Columbia University Press, New York

Pantoja, M. D. (2008). 'Acordos entre empresas e comunidades para o manejo florestal: uma análise crítica do caso das comunidades da Reserva Extrativista Rio Preto-Jacundá em Machadinho da Oeste, Rondônia, Brasil', MSc thesis, Núcleo de Ciências Agrárias e Desenvolvimento Rural (NAEA)/Universidad Federal do Pará (UFPA), Belém

Parker, N., Berthin, G., De Michele, R. and Mizrahi, Y. (2004). *Corruption in Latin America: A desk assessment. Americas' Accountability Anti-Corruption Project*, United States Agency for International Development (USAID), Washington D.C.

Peralta, C., Vos, V., Llanque, O. and Zonta, A. (eds) (2009). 'Productos del bosque: Potencial social, natural y financiero en hogares de pequeños productores de la Amazonia', Report of the ForLive Livelihood Workpackage 3, Universidad Autônoma do Beni «Jose Ballivián» (UAB), Riberalta, [online] http://www.waldbau.uni-freiburg. de/forlive/

Perz, S. (2005). 'The importance of household asset diversity for livelihood diversity and welfare among small farm colonists in the Amazon', *Journal of Development Studies,* vol 41, no 7, pp.1193–220

Perz, S., Walker, R. and Caldas, M. (2006). 'Beyond population and environment: household demographic life cycles and land use allocation among small farms in the Amazon', *Human Ecology,* vol 34, no 6, pp.829–49

Pfaff, A. S. P. (1999). 'What drives deforestation in the Brazilian Amazon? Evidence from satellite and socioeconomic data', *Journal of Environmental Economics and Management,* vol 37, pp.26–43

Pitman, C. A. N., Azaldegui, C. L. M., Salas, K., Vigo, G. and Lutz, A. D. (2007). 'Written accounts of an Amazonian landscape over the last 450 years', *Conservation Biology,* vol 21, no 1, pp.253–62

Pokorny, B. (2003). 'Forest management by small farmers in the Amazon – An opportunity to enhance forest ecosystem stability and rural livelihood', INCO Project Proposal (submitted 09.09.2003)

Pokorny, B., Cayres, G. and Nunes, W. (2005). 'Participatory extension as a basis for the work of rural extension services in the Amazon', *Agriculture and Human Values,* vol 22, no4, pp.435–50

Pokorny B. and Steinbrenner M. (2005). 'Collaborative monitoring of production and costs of timber harvest operations in the Brazilian Amazon', Ecology and Society, vol 10, no 1: 3, [online] http://www.ecologyandsociety.org/vol10/iss1/art3/

Pokorny, B. and Montero, I. (eds) (2007). *Deutschland und die Wälder Amazoniens,* University of Freiburg

Pokorny, B. and Phillip, M. (2008). 'Certification of NTFP. Concluding comments', *Forest, Trees and Livelihoods,* vol 18, no 1, pp.91–5

Pokorny, B. and Johnson, J. (2008a). 'Community forestry in the Amazon: The unsolved challenge of forests and the poor', ODI Natural Resource Perspectives, vol 112, Overseas Development Institute (ODI), London

—(2008b). 'Estrategias de acompañamiento al manejo forestal comunitario', in C. Sabogal, W. de Jong, B. Pokorny and B. Louman (eds) *Manejo forestal comunitario en América tropical: Experiencias, lecciones aprendidas y retos para el futuro,* Center for International Forestry Research (CIFOR), Bogor, pp.231–78

—(2008c). 'Forest research in development contexts. The concept of the accountable researcher', in International Union of Forest Research Organizations (IUFRO), *Forest research management in an era of globalization,* Proceedings of the IUFRO Unit 6.06.00, April 18–21, 2007, Arlington, Virginia, Society of American Foresters, Bethesda, pp.43–53

Pokorny, B., Sabogal, C., Stoian, D., De Jong, W., Louman, B., Pacheco, P. and Porro, N. (2009). 'Experiencias y retos del manejo forestal comunitario en América tropical', *Recursos Naturales y Ambiente,* vol 54, pp.81–98

Pokorny, B., Hoch, L. and Maturana, J. (2010a). 'Smallholder forest plantations in the tropics – Local people between outgrower schemes and reforestation programs', in J. Bauhus, P. van der Meer and M. Kanninen (eds) *Ecosystem goods and services from plantation forests,* Earthscan, London, pp.140–70

Pokorny, B., Sabogal, C., De Jong, W., Pacheco, P., Porro, N., Louman, B. and Stoian,

D. (2010b). 'Challenges of community forestry in tropical America', *Bois et Forêts des Tropiques*, vol 303, no 1, pp.43–66

Pokorny, B., Johnson, J., Medina, G. and Hoch, L. (2012). 'Market-based conservation of the Amazonian forests: Revisiting win–win expectations', *Geoforum*, vol 43, no 3, pp.387–401

Pokorny, B., Scholz, I. and de Jong, W. (in press a). 'REDD+ for the poor or the poor for REDD+? About the limitations of environmental sector policies and the potential of achieving environmental goals in the Amazon through consistent pro-poor policies', accepted by *Ecology and Society*

Pokorny, B., De Jong, W., Godar, J., Pacheco, P., and Johnson, J. (in press b). 'From large to small: reorienting development policies in response to climate change, food security and poverty', accepted by *Forest Policy and Economics*

Polanyi, M. (1983). *The tacit dimension*, Peter Smith, Gloucester

Porro, N., Germana, C., López, C., Medina, G., Ramírez, Y., Amaral, M. and Amaral, P. (2008). 'Capacidades organizativas para el manejo forestal comunitario frente a las demandas y expectativas oficiales', in C. Sabogal, W. de Jong, B. Pokorny and B. Louman (eds) *Manejo forestal comunitario en América tropical: Experiencias, lecciones aprendidas y retos para el futuro*, Center for International Forestry Research (CIFOR), Bogor, pp.163–94

Porter-Bolland, L., Ellis, A. E., Guariguata, R. M., Ruiz-Mallén, I., Negrete-Yankelevich, S. and Reyes-García, V. (2012). 'Community managed forests and forest protected areas: An assessment of their conservation effectiveness across the tropics', *Forest Ecology and Management*, vol 268, no 1, pp.6–17

Prabhu, R., Colfer, C. and Dudley, R. (1999). 'Guidelines for developing, testing and selecting criteria and indicators for sustainable forest management', The Criteria and Indicators Toolbox Series 1, Centre for International Forestry Research (CIFOR), Bogor

Putz, F. E., Zuidema. P. A., Pinard, M. A., Boot, R., Sayer. J. et al. (2008). 'Improved tropical forest management for carbon retention', *PLoS Biol*, vol 6, no 7, e166, [online] doi:10.1371/journal.pbio.0060166

Reardon, T., Delgado, C. and Matlon, P. (1992). 'Determinants and effects of income diversification amongst farm households in Burkina Faso', *Journal of Development Studies*, vol 28, pp.264–96

Rede Geoma (2004). *Dinâmica territorial da frente de ocupação de São Félix do Xingu-Iriri. Subsídios para o desenho de políticas emergenciais de contenção do desmatamento*, Museu Parense Emilio Goeldi (MPEG), Belém, Brazil

Reimberg, M. (2009). 'Municípios devastadores apresentam vínculo com escravidão', *Repórter Brasil*, 14 April, [online] www.amazonia.org.br/noticias/noticia.cfm?id=307288

Reis, A. (1997). *Seringal e o seringueiro*, 2nd edn, Editora da Universidad do Amazonas, Manaus

Ribot, J. and Peluso, L. (2003). 'A theory of access', *Rural Sociology*, vol 68, no 2, pp.153–81

Robles, M. (in preparation). 'Potential of forests for smallholders in the Amazon', PhD thesis, University of Freiburg

Rocheleau, E. D. (2008). 'Political ecology in the key of policy: From chains of explanation to webs of relation', *Geoforum*, vol 39, pp.716–27

Rodrígues, A., Ewers, R. M., Parry, L., Souza Júnior, C., Veríssimo, A. and Balmford, A. (2009). 'Boom-and-bust development patterns across the Amazon deforestation frontier', *Science*, vol 324, no 5933, pp.1435–7

Rodríguez, A. (2008). *El uso de tecnologías intermedias como aporte de valor agregado para la producción forestal y ventaja competitiva comercial del Manejo Forestal Comunitario*. Instituto para el Hombre Agricultura y Ecología (IPHAE), Riberalta

Rodrigues, R. L. V. (2004). 'Análise dos fatores determinantes do desflorestamento na Amazônia legal', PhD thesis, Universidad Federal de Rio de Janeiro (UFRJ)

Rogers, E. M. (2003). *Diffusion of innovations*, 5th edn, The Free Press, New York

Romero, D. (2011) 'The challenge of implementing good ideas: a case of study from the forest sector in Bolívia', MSc thesis, University of Freiburg

Rose-Ackerman, S. (2007). *Public administration and institutions in Latin America*. Copenhagen Consensus, San José de Costa Rica

Rostow, W. W. (1960). *The stages of economic growth: A non-communist manifest*, Cambridge University Press, Cambridge

Rudel, T. K., Defries, R., Asner, G. P. and Laurance, W. F. (2009). 'Changing drivers of deforestation and new opportunities for conservation', *Conservation Biology*, vol 23, pp.1396–405

Ruiz, S. A. (2005) 'Institutional change and social conflicts over forest use in the northern Bolivian Amazon', Freiburger Schriften zur Forst-und Umweltpolitik Band 10, Dr. Kessel, Remagen-Oberwinter, Germany

Ryder, R. and Lawrence, A. B. (2000). 'Urban-system evolution on the frontier of the Ecuadorian Amazon', *Geographical Review*, vol 90, no 4, pp.511–35

Sabogal, C., De Jong, W., Pokorny, B. and Louman, B. (eds) (2008b). *Manejo forestal comunitario en América tropical: Experiencias, lecciones aprendidas y retos para el futuro*. Centre for International Forestry Research (CIFOR), Bogor

Sabogal, C., Nalvarte, J. and Colán, V. (2008a). *Análisis del marco legal para el manejo forestal por pequeños productores y comunidades en la Amazonia Peruana*. Asociación para la Investigación y el Desarrollo Integral (AIDER), Lima

Sabogal, C., Snook, L., Boscolo, M., Pokorny, B., Quevedo, L., Lentini, M. and Colán, V. (2007). 'Adopción de prácticas de manejo forestal sostenible por empresas madereras', *Recursos Naturales y Ambientales*, vol 49, pp.100–11

Salgado, I. and Kaimowitz, D. (2003). 'Porto de Moz: o prefeito, "dono do município"', in F. Toni and D. Kaimowitz (eds) *Municípios e gestão florestal na Amazônia*, Editores A. S., Natal, pp.219–52

Sanchez, M. (2007). 'Latin America – the 'Persian Gulf ' of biofuels?', *The Washington Post*, 23 February, [online] www.washingtonpost.com/wp-dyn/content/article/2007/02/22/AR2007022201361.html

Santana, A. C., Gomes, S.,C., Fernandes, A. R. and Botehlo, M. N. (2003). *Perfil do profissional de ciências agrárias formado na Universidade Federal Rural da Amazônia: empregadores, graduados e instituições correlatas,* Universidade Federal Rural da Amazônia (UFRA), Belém, Brazil

Sato, R. (2010). 'Local benefits from payments for environmental services: exploring the case of bolsa floresta in the Brazilian Amazon', MSc thesis, University of Freiburg

Sato, R., Steinbrenner, M. and Pokorny, B. (in preparation). 'Acordos informais para a exploração de Madeira em área de pequenos produtores: revisão de preconceitos diante do manejo florestal comunitário', in B. Pokorny, I. Montero, J. C. Montero and J. Johnson (eds) *Uso forestal por pequeños productores en la Amazonía: En busca de evidencias empíricas para los grandes paradigmas*, University of Freiburg

Sayer, J. (1995). 'Science and international nature conservation', Occasional Paper, vol 4, Centre for International Forestry Research (CIFOR), Bogor

Sayer, J. and Campbell, B. (2001). 'Research to integrate productivity enhancement, environmental protection, and human development', *Conservation Ecology*, vol 5, no 2: 32, [online] http://www.consecol.org/vol5/iss2/art32

Scherr, S., White, A., and Kaimowitz, D. (2001). *Making markets work for forest communities*, Forest Trends, Washington D.C.

Schmink, M. (1984). 'Household economic strategies: review and research agenda', *Latin American Research Review*, vol 19, pp.87–101

Schmink, M. and Wood, C. (1992). *Contested frontiers in the Amazon*, Columbia University Press, New York

Schneider, S. (2003). 'Desenvolvimento rural regional e articulações extra-regionais' in Proceedings of the International Forum *Território, desenvolvimento rural e democracia*, in Fortaleza, 16–19 November, pp.1–23

Schulze, M. D., Grogan, J., Uhl, C., Lentini, M. and Vidal, E. (2008) '.Evaluating ipê *(Tabebuia, Bignoniaceae)* logging in Amazonia: sustainable management or catalyst for forest degradation?' *Biological Conservation*, vol 141, pp.2071–85

Scoones, I. (1998). 'Sustainable rural livelihoods: a framework for analysis', Working Paper, vol 72, Institute of Development Studies (IDS), Brighton

Scott, A. J. and Garofoli, G. (2007). *Development on the ground*, Routledge, London

Scott, W. R. (2001). *Institutions and organizations*. 2nd edn, Sage Publications, London

Selener, D. (1997). *Farmer-to-farmer extension: Lessons from the field*, Instituto Internacional de Reconstrucción Rural (IIRR), Quito

Shackleton, S., Campbell, B, Lotz-Sisitka, H. and Shackleton, C. (2008). 'Links between the local trade in natural products, livelihoods and poverty alleviation in a semi-arid region of South Africa', *World development*, vol 36, pp.505–26

Shanley, P., Pierce, A. R., Laird, S. A. and Guillén, A. (eds) (2002). *Tapping the green market: Certification and management of non-timber forest products*, Earthscan, London

Shanley, P., Pierce, A., Laird, S. and Robinson, D. (2008). *Beyond timber: Certification and management of non-timber forest products*, Centre for International Forestry Research (CIFOR), Bogor

Shanley, P. and Rosa, N. A. (2004). 'Eroding knowledge: An ethnobotanical inventory in Eastern Amazonia's logging frontier', *Economic Botany*, vol 58, no 2, pp.135–60

Sheil, D., Puri, R. K., Basuki, I., Van Heist, M., Syaefuddin, R., Agung Sardjono, M. A., Samsoedin, I., Sidiyasa, K., Chrisandini, Permana, E., Mangopo Angi, E., Gatzweiler, F. Johnson, B. and Wijaya, A. (2003). *Exploring biological diversity, environment and local people's perspectives in forest landscapes. Methods for a multidisciplinary landscape assessment*, 2nd edn, Centre for International Forestry Research (CIFOR), Bogor

Shepherd, A. W. (2007), 'Approaches to linking producers to markets', Agricultural Management, Marketing and Finance Occasional Paper, vol 13, FAO, Rome

Sikor, T. and Stahl, J. (eds), (2011) *Forests and people: Property, governance, and human rights*, Routledge, London

Silva, E., Kaimowitz, D., Bojanic, A., Ekoko, F., Manurung, T. and Pavez, I. (2002), 'Making the law of the jungle: The reform of forest legislation in Bolivia, Cameroon, Costa Rica, and Indonesia', *Global Environmental Politics*, vol 2, no 3, pp.63–97

Silva, J. N. M., Carvalho, J. O. P., Lopes, J. C. A., Almeida, B. F., Costa, D. H. M., Oliveira, L. C., Vanclay, J. K. and Skovsgaard, J. P. (1995), 'Growth and yield of a tropical rain forest in the Brazilian Amazon 13 years after logging', *Forest Ecology and Management*, vol 71, no 3, pp.267–274

Simmons, S., Walker, R. T. and Wood, C. H. (2002). 'Tree planting by small producers in the tropics: A comparative study of Brazil and Panama', *Agroforestry Systems*, vol 56, no 2, pp.89–105

Skole, D. and Tucker, C. (1993). 'Tropical deforestation and habitat fragmentation in the Amazon: satellite data from 1978 to 1988', *Science*, vol 260, no 5116, pp.1905–10

Smeraldi, R. and May, P. (2008). *The cattle realm: a new phase in the livestock colonization of Brazilian Amazonia*, Amigos da Terra, Rio de Janeiro

Smith, J., Sabogal, C., De Jong, W. and Kaimowitz, D. (1997). 'Bosques secundários como recurso para el desarollo rural y la conservación ambiental en los trópicos de América latina', CIFOR Occasional Paper, vol 13, Centre for International Forestry Research (CIFOR), Bogor

Soares, L. (2008) *Os efeitos da demanda crescente de produtos extrativos para os pequenos produtores de açaí (Euterpe oleracea Mart.) na microregião de Cametá – Pará.* Núcleo de Ciências Agrárias e Desenvolvimento Rural (NAEA)/Universidad Federal do Pará (UFPA), Belém, Brazil

Soares-Filho, B. S., Nepstad, D., Curran, L. M., Cerqueira, G. C., Garcia, R. A., Azevedo Ramos, C., Voll, E., Macdonald, A., Lefebvre, P. and Schlesinger, P. (2006). 'Modelling conservation in the Amazon basin', *Nature*, vol 440, pp.520–3

Sobral Escada, M. I., Vieira, I. and Kampel, S. A. (2005). 'Processos de ocupação nas novas fronteiras da Amazônia (o interflúvio do Xingu/Iriri)', Estudos Avancados, vol 19, pp.9–23

Sombroek, W. and Carvalho de Souza, A. (2002). 'Macro- and micro ecological–economic zoning in the Amazon region history, first results, lessons learnt and research needs', in R. Lieberei, R. Bianchi, H. K. Boehm and V. Reisdorff (eds) Neotropical Ecosystems, Proceedings, German–Brazilian Workshop, GKSS-Geesthacht (Hamburg, 2000), pp91–98, [online] http://www.biologie.uni-hamburg.de/bzf/oknu/proceedingsneotro-pecosys/p0091-p0098_reflections_macro_and_micro.pdf

Spang, L. (2005). 'IIRSA Update #2', Bank Information Centre, Washington D.C., 3 June [online] http://www.bicusa.org/en/Article.2139.aspx

Stiglitz, J. E. (2006). *Livre mercado para todos'*, Campus Editora, São Paulo

Steinbrenner, R.A. (2011). *Mídia alternativa para um desenvolvimento alternativo na Amazônia*, PhD thesis, Núcleo de Ciências Agrárias e Desenvolvimento Rural (NAEA)/Universidade Federal do Pará (UFPA), Belém, Brazil

Stöhr, W. B. (ed.) (1990). *Global challenge and local response*, Mansell, London

Stoian, D. (2005). *La economía extractivista de la Amazonía norte boliviana.* Centre for International Forestry Research (CIFOR), Bogor

Stoian, D. and Donovan, J. (2004). 'Articulación del mundo campesino con el mercado: Integración de los enfoques de medios de vida y cadena productiva', in Centro Agronómico Tropical de Investigación e Enseñanza (CATIE) (ed) *Memorias de la semana científica*. CATIE, Turrialba, Costa Rica, pp.14–16

Sunderlin, W. D., Angelsen, A., Belcher, B., Burgers, P., Nasi, R., Santoso, L. and Wunder, S. (2005). 'Livelihoods, forests and conservation in developing countries: An overview', *World Development*, vol 33, no 9, pp.1383–402

Sunderlin, W. D., Hatcher, J. and Liddle, M. (2008). *From exclusion to ownership? Challenges and opportunities in advancing forest tenure reform*, Rights and Resource Institute (RRI), Washington D.C.

Swift, J. (1998). *Factors influencing the dynamics of livelihood diversification and rural non-farm employment in space and time. Rural non-farm employment project.* Natural Resources Institute, Chatham

Tacconi, L. (2007). Illegal logging: Law enforcement, livelihoods and the timber trade. Earthscan, London

Tauk Santos, M. S. (1998). 'Gestão da comunicação no desenvolvimento regional', Comunicação e Educação, vol 11, Editora Moderna, São Paulo

—(2000). 'Comunicação rural e mercado de trabalho na era tecnológica: o desenvolvimento local está em pauta', in A. B. F. Callou (ed.) *Comunicação rural e o novo espaço agrário*, Fractais, vol 3, Universidade Federal Rural de Pernambuco (UFRPE), Recife

TEBTEBBA (Indigenous Peoples' International Centre for Policy Research and Education) (2008). 'Global indigenous peoples consultation on reducing emissions from deforestation and forest degradation (REDD)', Summary Report of conference held in Baguio City, Philippines, 12–14 November

Terluin I. J. (2003). 'Differences in economic development in rural regions of advanced countries: an overview and critical analysis of theories', *Journal of Rural Studies*, vol 19, pp.327–44

Teubal, M. (2009). 'Agrarian reform and social movements in the age of globalization. Latin America at the dawn of the twenty-first century', *Latin American Perspectives*, vol 26, pp.9–20

Texeira, E. S., Bernartt, L. M. and Trindade, G. A. (2008). 'Estudos sobre a pedagogia da alternância no Brasil: revisão de literatura e perspectivas para a pesquisa', *Revista Educação e Pesquisa*, vol 34, no 2, pp.227–42

Thomas, J. (2008). 'REDD governance: corruption as a catalyst for deforestation in Ecuador', MSc thesis, University of Freiburg

Treccani, G. D. (2008). *Estado do Pará: do caos à terra de direitos (Campo Cidadão)*, Instituto de Terras do Pará (ITERPA), Belém, Brazil

Trejo, K. (2007). 'Decisive life events. Small farmers coping and adapting. – The Bolivian Amazon case', MSc thesis. University of Freiburg

Tucker, C. J. and Townshend J. R. G. (2000). 'Strategies for monitoring tropical deforestation using satellite data', *International Journal of Remote Sensing*, vol 21, no 6–7, pp.1461–71

UNCED (United Nations World Conference on Environment and Development) (1987). *Our common future* (Brundtland Report), Oxford University Press, Oxford

—(1992). 'Agenda 21: Deforestation', Report of the UNCED, Rio de Janeiro, 3–14 June 1992, United Nations, New York

UNDP (United Nations Development Programme) (1997). 'Manejo, conservación y utilización de los recursos forestales en el trópico de Cochabamba y en las zonas de transición de los Yungas de La Paz', Report Phase II, Project: AD/BOL/97/C23

—(2010). *Regional human development report for Latin America and Caribbean 2010: acting on the future breaking the intergene-racional transmisión of inequality*, UNDP, San José

UNFCC (United Nations Framework Convention on Climate Change) (2011). 'Framework convention on climate change', Report of the Conference of the Parties on its 16th session held in Cancun 29 November–10 December 2010, FCCC/CP/2010/7, United Nations, New York

Vázquez-Barquero, A. (1993). *Política económica local*, Editorial Pirámide, Madrid

—(2002). *Endogenous development. Networking, innovation, institutions and cities*, Routledge, London

—(2010). *The new forces of development: Territorial policy for endogenous development*. World Scientific Publishing Company, Singapore

Vázques-Barquero, A. and Garofoli, F. (1995). *Desarollo económico rural en Europa*, Colegio de Economistas de Madrid, Madrid

Vedeld, P., Angelsen, A., Sjaastad, E., and Kobugabe-Berg, G. (2004). 'Counting on the environment. Forest incomes and the rural poor', Environmental Economics Series 98, World Bank, Washington D.C.

Vieira, I., De Almeida, A. S., Davidson, E. A., Stone, T. A., De Carvalho, C. J. R. and Guerrero, J. B. (2003). 'Classifying successional forests using Landsat spectral properties and ecological characteristics in eastern Amazonia', *Remote Sensing of Environment*, vol 87, no 4, pp.470–81

Villacis, E. (2010). 'Why producers do not participate in organic certified agriculture? The case of small farmers in the Amazon Highway, Brazil', MSc thesis, University of Freiburg

Vos, V., Llanque, O. and Zonta, A. (eds) (2009). 'Medios de vida y manejo forestal por pequeños productores de la Amazônia. Report of the ForLive Livelihood Workpackage 3. Universidad Autônoma do Beni „Jose Ballivián" (UAB), Riberalta, [online] http://www.waldbau.uni-freiburg.de/forlive/

Wagner, B. A. A. (2008). *Terra de quilombo, terras indígenas, "babaçuais livre", "castanhais do povo", faixinais e fundos de pasto: terras tradicionalmente ocupadas.* 2nd edn, Universidade Federal do Amazonas (UFAM), Manaus

Ward, J. H. (1963). 'Hierarchical grouping to optimize an objective function', *Journal of the American Statistical Association*, vol 58, pp.236–44

Wehrlich D. P. (1978). *Peru: A Short History*, Southern Illinois University Press, Carbondale

Weigelt, J. (2011). 'Reforming development trajectories? Institutional change of forest tenure in the Brazilian Amazon', PhD thesis, Humboldt University Berlin

Welch, C. (2009). 'Camponeses: Brazil's peasant movement in historical perspective (1946–2004)', *Latin American Perspectives*, vol 36, no 4, pp.126–55

White, A. and Martin, A. (2002). *Who owns the world's forests?*, Forest Trends, Washington D.C.

Wienold, H. (2006).'Brasiliens Agrarfront: Landnahme, Inwertsetzung und Gewalt', *Peripherie*, vol 101/102, pp.43–68

Wiersum, F. (2004). 'Forest gardens as an intermediate land use system in the nature-culture continuum: characteristics and future potential', *Agroforestry Systems*, vol 61, pp.123–34

Wiggins, S, Kirsten, J. and Llambi, L. (eds) (2010). 'The future of small farms', *World Development*, vol 38, no 10, pp.1341–526

World Bank (2007). '*World development report 2008. Agriculture for development*', World Bank, Washington D.C.

—(2008). 'Social dimensions of climate change', Workshop Report, World Bank, Washington D.C.

Wunder, S. (2001). 'Poverty alleviation and tropical forests – What scope for synergies', *World Development*, vol 19, no 11, pp.1817–33

—(2005). 'Payments for environmental services: Some nuts and bolts', CIFOR Occasional Paper, vol 42, Centre for International Forestry Research (CIFOR), Bogor

Yin, R. (2009). *Case study research: design and methods.* 4th edn, SAGE Publications, Thousand Oaks, USA

Zoomers, A. (ed.) (2001). *Land and sustainable livelihood in Latin America.* KIT Publications, Amsterdam

Index

For Product Safety Concerns and Information please contact our EU
representative GPSR@taylorandfrancis.com
Taylor & Francis Verlag GmbH, Kaufingerstraße 24, 80331 München, Germany

www.ingramcontent.com/pod-product-compliance
Lightning Source LLC
Chambersburg PA
CBHW050423280326
41932CB00013BA/1972